OECD Companion to the Inventory of Support Measures for Fossil Fuels 2015

This work is published under the responsibility of the Secretary-General of the OECD. The opinions expressed and arguments employed herein do not necessarily reflect the official views of OECD member countries.

This document and any map included herein are without prejudice to the status of or sovereignty over any territory, to the delimitation of international frontiers and boundaries and to the name of any territory, city or area.

Please cite this publication as:
OECD (2015), *OECD Companion to the Inventory of Support Measures for Fossil Fuels 2015*, OECD Publishing, Paris.
http://dx.doi.org/10.1787/9789264239616-en

ISBN 978-92-64-23960-9 (print)
ISBN 978-92-64-23961-6 (PDF)

The statistical data for Israel are supplied by and under the responsibility of the relevant Israeli authorities. The use of such data by the OECD is without prejudice to the status of the Golan Heights, East Jerusalem and Israeli settlements in the West Bank under the terms of international law.

Photo credits: Cover © klikk - Fotolia.com, © iamtheking33 - Fotolia.com, © umabatata - Fotolia.com, © Anzelm - Fotolia.com, © Ghost - Fotolia.com

Corrigenda to OECD publications may be found on line at: *www.oecd.org/about/publishing/corrigenda.htm*.

Foreword

The year 2015 should prove to be a momentous one in the battle against climate change. Global leaders and policymakers will meet in Paris in December with a social, economic, political, environmental, and moral imperative to reach an ambitious and actionable agreement at COP 21. Meeting the challenge of climate change requires us to achieve zero net emissions from fossil fuels globally by the end of this century. Without zero net CO_2 emissions, temperatures will just keep rising.

Central to tackling climate change must be concerted efforts to reform "lose-lose" fossil-fuel subsidies. By distorting costs and prices, fossil-fuel subsidies create inefficiencies in the way we generate and use energy. They are also costly for governments, crowding out scarce fiscal resources that could be put to better use, such as strategic investment in the education, skills, and physical infrastructure that people most value in the 21st century. But most importantly, fossil-fuel subsidies undermine efforts to make our economies less carbon-intensive while exacerbating the damage to human health caused by air pollution.

Increasing interest in the reform of fossil-fuel subsidies has been most evident in the context of the G20 and APEC, with several governments volunteering over the last two years to be peer-reviewed on some of their policies supporting fossil fuels. Likewise, the recent steps taken by several OECD Key Partner economies — India and Indonesia, first and foremost — are proof that subsidy reform pays off. Those are undoubtedly important achievements, though more needs to be done. There is also an enduring need to ensure that the poorest members of society are not adversely affected in the process.

A key lesson from OECD work on measures supporting fossil fuels is that transparency matters. Not only do citizens need to understand how the taxes they pay are spent, but the sharing of experiences among countries is also crucial in spreading best practices and sound policy. This book and its associated database do just that. By identifying and documenting almost 800 individual policies that support the extraction, refining, or combustion of fossil fuels in OECD countries and large emerging economies, they highlight the need for governments to periodically review their budgets and tax codes in light of changing circumstances and evolving policy priorities.

The value of all measures contained in the OECD database amounted to between USD 160-200 billion annually over the period 2010-14. Not all these policies are unambiguously inefficient, and some caution is required in interpreting these amounts. Nevertheless, there is clearly ample scope to save scarce budgetary resources and improve the environment in both advanced and emerging economies. As governments progress in their efforts to rationalise and phase out policies supporting fossil fuels, the OECD stands ready to help by exploring more efficient and effective "win-win" alternatives to achieve *better policies for better lives*.

Angel Gurría
Secretary-General

Acknowledgments

This report was prepared by Jehan Sauvage of the OECD's Trade and Agriculture Directorate under the supervision of Franck Jésus and Ronald Steenblik. Special thanks go to Theresa Poincet for her help in formatting early versions of the document, and Michèle Patterson for the final preparation of the manuscript for publication. Excellent statistical and research assistance was provided by Stefanie Heerwig, Mark Mateo, and Ella Rebalski. Helpful comments and suggestions were also provided by a number of colleagues at the OECD and the IEA: Johanna Arlinghaus, Simon Bennett, Gregory Briner, Amos Bromhead, Romain Champetier, Florens Flues, Michelle Harding, Florian Kitt, Nora Selmet, Janine Treves, Kurt van Dender, and Georgios Zazias. The help of Christophe de Gouvello and Masami Kojima from the World Bank is gratefully acknowledged too.

Several individuals contributed to the expansion and update of the database (the Inventory) and merit acknowledging, starting with Caroline Gomes Nogueira, Stefanie Heerwig, Mark Mateo, and Ella Rebalski, who played decisive roles at various stages of the project. A number of other people helped collect and update information for particular countries, including: Kaushik Bandyopadhyay (India), Anomitro Chatterjee (India), Filippo Civitelli (Italy), Evelyne Ferraris (Brazil), Ivetta Gerasimchuk (the Russian Federation), Robert T. Klein (the People's Republic of China), George Mergos (Greece), Suja Moon (Korea), Chikara Onda (Japan), Kateryna Perepechay (the Russian Federation), Laszlo Pinter (Hungary), Ronny Regev (Israel), Jocelyn van Berkel (Netherlands), and Chan Yang (the People's Republic of China). The contributions of Éric Espinasse, Frano Ilicic, and Nobuko Miyachiyo are also gratefully acknowledged for the important role they played in developing the online database.

This report and the associated database were discussed at various stages of their development by the OECD's Joint Meetings of Tax and Environment Experts. They were also discussed at, and approved for publication by, the Committee on Fiscal Affairs and the Environment Policy Committee. Invaluable information, comments, and other input concerning the report and the associated database were provided by Delegates to the Joint Meetings and their colleagues in national and sub-national government administrations.

Table of contents

Tables

Figures

Boxes

Acronyms

ADB	Asian Development Bank
APEC	Asia-Pacific Economic Cooperation
CCS	Carbon capture and storage
CGE	Computable general equilibrium
CNG	Compressed natural gas
CO_2	Carbon dioxide
CO_{2eq}	Carbon dioxide equivalent
CSE	Consumer Support Estimate
EHS	Environmentally harmful subsidy
END	Other end uses
EREP	European Resource Efficiency Platform
ETS	Emissions trading system
EXTRACT	Exploration and extraction
FFFSR	Friends of Fossil Fuel Subsidy Reform
GENER	Power and heat generation
GHG	Greenhouse-gas
GJ	Gigajoule
GSSE	General Services Support Estimate
Gt	Gigatonne
HS	Harmonized System
IADB	Inter-American Development Bank
IEPS	Impuesto Especial sobre Producción y Servicios
IMF	International Monetary Fund
INDUS	Industrial processes and activities
IPCC	Intergovernmental Panel on Climate Change
IRS	Internal Revenue Service
LNG	Liquefied natural gas
LPG	Liquefied petroleum gas
MFN	Most-favoured nation
MW	Megawatt
NO_X	Nitrogen oxides
O_3	Ozone
OMB	Office of Management and Budget
PM	Particulate matter
PSE	Producer Support Estimate
REFIN	Refining and processing
SCM	Agreement on Subsidies and Countervailing Measures
SDGs	Sustainable Development Goals
SO_2	Sulphur dioxide
TRANS	Bulk transportation and storage
US EIA	United States Energy Information Administration
VAT	Value-added tax
VOCs	Volatile organic compounds
WEO	World Energy Outlook
WTO	World Trade Organization

Executive summary

The combustion of fossil fuels is a leading contributor to climate change, and many countries have already taken steps to reduce their emissions of CO_2 and other pollutants. Some policies remain, however, that encourage more production and use of fossil fuels than would otherwise be the case. In so doing, these policies increase emissions and make mitigation more costly than necessary. Fossil-fuel subsidies are one such policy.

Not only do fossil-fuel subsidies undermine efforts to mitigate climate change, but they are also costly and distortive. By distorting costs and prices, fossil-fuel subsidies generate inefficiencies in the production and use of energy economy-wide. This can affect the allocation of resources across industries, including by directing long-term capital investment toward sectors that produce fossil fuels or use them intensively, at the expense of cleaner forms of energy and other economic activities more generally. In so doing, these subsidies accentuate the risk that long-lived capital assets end up locking in polluting technologies for years or decades. Fossil-fuel subsidies can also impose considerable strain on government budgets since they either increase public expenditure or reduce tax revenue, and this at a time when many countries are taking painful steps to reduce their public debt.

What really differentiate fossil-fuel subsidies from most other types of subsidies, however, are their environmental impacts. Besides greenhouse-gas emissions, the extraction of fossil fuels and their combustion in power plants, vehicles, and buildings, are directly responsible for the emission of numerous pollutants having local and often immediate impacts on the environment and on human health. This imposes additional costs on society where governments do not take appropriate action to ensure that polluters pay. To the extent they increase the production or consumption of fossil fuels, fossil-fuel subsidies make matters worse by indirectly rewarding polluting behaviour.

For all these reasons, a number of international initiatives over the past years have called for the reform of fossil-fuel subsidies. Some, like the commitments by APEC and the G-20, involve broad sets of countries while others remain more regional in scope, such as the efforts undertaken by the European Commission and certain regional development banks. All these initiatives proceed from the perspective that fossil-fuel subsidies are eminently bad and that reforming them requires some degree of international co-operation.

To help improve understanding of the range and magnitude of fossil-fuel subsidies, the OECD has identified, documented, and estimated almost 800 individual policies that support the production or consumption of fossil fuels in OECD countries and six large partner economies (Brazil, the People's Republic of China, India, Indonesia, the Russian Federation, and South Africa). In line with previous OECD work looking at support for the agriculture sector and the fisheries sector, the scope of the policies inventoried here is broad and differs from some conceptions of "subsidy". It includes both direct budgetary transfers and tax expenditures that in some way provide a benefit or preference for fossil-fuel production or consumption relative to alternatives.

The database and the present report do not provide any analysis of the impacts of specific support measures, and so do not pass any judgement on which measures might be usefully kept in place, and which ones a country might wish to consider for possible reform or removal. Their purpose is rather to provide comprehensive information about policies that confer some level of support, as a starting point for subsequent analysis looking at the objectives of particular measures, their impacts (economically, environmentally and socially), and possible reforms and alternatives. First and foremost, this inventory seeks to promote the transparency of public policies and government budgets, and eventually a greater accountability for how public resources are used. It can also be understood as a contribution toward the broader issue of how to make fiscal policy and tax systems "greener" or more environmentally friendly.

Using data obtained from government sources, the report finds that the many measures the database contains had an overall value of USD 160-200 billion annually over the period 2010-14, with support for the consumption of petroleum products accounting for the bulk of that amount. This reflects in part the importance of petroleum products in countries' total primary energy supply, but also owes much to the fact that many large OECD economies do not extract fossil fuels on a significant scale. Producer support is unsurprisingly much more significant in relative terms when looking at countries that are relatively well endowed with crude oil, natural gas or coal (e.g. Canada, Germany, the Russian Federation or the United States).

Compared with the previous edition of the *Inventory* (OECD, 2013b), which covered OECD countries only, support now seems to follow a downward trend after having peaked twice in 2008 and 2011-12. Although the decline is more marked in OECD countries, a similar downward trend can also be observed in partner economies, driven in part by India's recent reform of its subsidies for the consumption of diesel fuel. A sizable portion of the decrease in support observed for OECD countries can be ascribed to Mexico, which eliminated the support it provided for the consumption of gasoline and diesel fuel through its floating excise tax.

The results reveal a certain degree of inertia, as policies supporting fossil fuels tend to stay in place for protracted periods of time. Most measures (about two-thirds of them) seem to have been introduced prior to 2000, at a time when climate change was not deemed a concern among policy makers and political and economic circumstances were by and large different. What this suggests is that there might be a need for countries to periodically reassess the relevance of some of their support measures as the context evolves.

Although progress has been notable, this edition of the *Inventory* shows that there remains plenty of room for reform. The time is also not one for complacency. Global greenhouse-gas emissions are still largely above the levels required for limiting average expected temperature increases. Recovery from the Great Recession of 2008-09 remains slow and difficult by historical standards. Fiscal positions continue posing a challenge to policy makers in many countries as they struggle to identify opportunities for cutting spending and raising more revenues, and this without adding to alarmingly high levels of unemployment. In this context, the reform of measures supporting fossil fuels appears more relevant than ever. Other better-targeted policy instruments likely exist that would offer suitable alternatives for meeting the policy objectives support measures for fossil fuels intended to reach in the first place.

Chapter 1

The case for measuring support for fossil fuels

The discussion in this first chapter sets the stage for better understanding the present report and the associated database. To do this, Section 1.1 looks at the reasons why fossil-fuel subsidies are generally considered to be harmful for the economy and the environment. Section 1.2 then shows how this helps explain the recent emergence of a consensus for reforming fossil-fuel subsidies, and how this growing consensus has led to a number of policy initiatives internationally and domestically. Section 1.3 then concludes by placing the OECD Inventory onto that broader stage, emphasising the important role it plays in ongoing discussions of energy policies and their reform.

The statistical data for Israel are supplied by and under the responsibility of the relevant Israeli authorities. The use of such data by the OECD is without prejudice to the status of the Golan Heights, East Jerusalem and Israeli settlements in the West Bank under the terms of international law.

1.1. Why reforming support for fossil fuels makes sense

Mitigating climate change requires coherent policy signals

As global greenhouse-gas (GHG) emissions continue to increase, so does the threat of higher average temperatures and their consequences for the environment and human welfare. A recent assessment by the Intergovernmental Panel on Climate Change (IPCC) shows these consequences to include unprecedented increases in sea levels, biodiversity loss, and a higher frequency of extreme weather events such as floods and droughts (IPCC, 2014). The assessment also indicates that emissions from fossil-fuel combustion, cement production, and flaring account for the majority of all anthropogenic emissions of CO_2 (Figure 1), a gas responsible for about two-thirds of all anthropogenic GHG emissions in 2010. This makes the use of fossil fuels a leading contributor to global warming and its effects on the natural environment.

Under such conditions, additional mitigation efforts will imply that significant action be taken to reduce countries' reliance on fossil fuels. Substantial progress has already been made to curb GHG emissions from fossil-fuel combustion in a number of countries, including large emitters like the People's Republic of China,[1] the European Union or the United States. Current efforts are, however, unlikely to be enough to avoid average temperatures increasing by 4°C above pre-industrial levels by 2100. The IPCC's latest assessment finds, for example, that scenarios in which the average temperature increases by less than 2°C over the 21st century involve global GHG emission reductions between 40% and 70% by 2050 compared with 2010 levels (IPCC, 2014). To achieve emission reductions on that scale would necessitate the widespread replacement of fossil-fuel-based energy sources with low-carbon energy sources and the deployment of technologies for capturing and storing CO_2.

Achieving the large-scale abatement of GHG emissions poses a number of challenges for policy makers. First among these are the short-term costs that mitigation measures may impose on economic actors, whether firms or households. Recent evidence shows these costs to vary widely depending on which policy instruments are used, with taxes and emission permits usually resulting in lower abatement costs (OECD, 2013a).[2] Reducing these costs further requires that signals and incentives for mitigation be coherent across policy areas, so that one set of policies does not undermine what the other is trying to achieve. In that regard, addressing climate change is as much about introducing new policies as it is about adapting existing ones. This need to better align policies across domains of public action was recently highlighted in an OECD report — *Aligning Policies for a Low-carbon Economy* — that identifies instances in which policy misalignments can hinder the effectiveness of low-carbon policies (OECD, 2015a).

Measures that directly support the production or unabated consumption of fossil fuels are prime examples of policies that run counter to mitigation objectives. Because they reduce the effective price of carbon, fossil-fuel subsidies make it more difficult to operate the necessary shift toward low-carbon energy sources. In that sense, they belong to the broader set of environmentally harmful subsidies (EHS), which have already been the object of several studies in the OECD context (e.g. OECD, 2003). The Organisation's interest in EHS goes back to Objective 1 of the *OECD Environmental Strategy for the First Decade of the 21st Century* that was adopted by OECD Environment Ministers in May 2001, and which already stressed the need to "remove or reform subsidies and other policies that encourage unsustainable use of natural resources — beginning with the agriculture, transport and energy sectors" (OECD, 2001). Objective 2 of the statement also emphasises the importance of "green tax reform", which is of particular importance for tax concessions encouraging the production and use of fossil fuels.

Figure 1.1. The combustion of fossil fuels has been the main contributor to global anthropogenic emissions of CO_2

Gt CO_2 per year

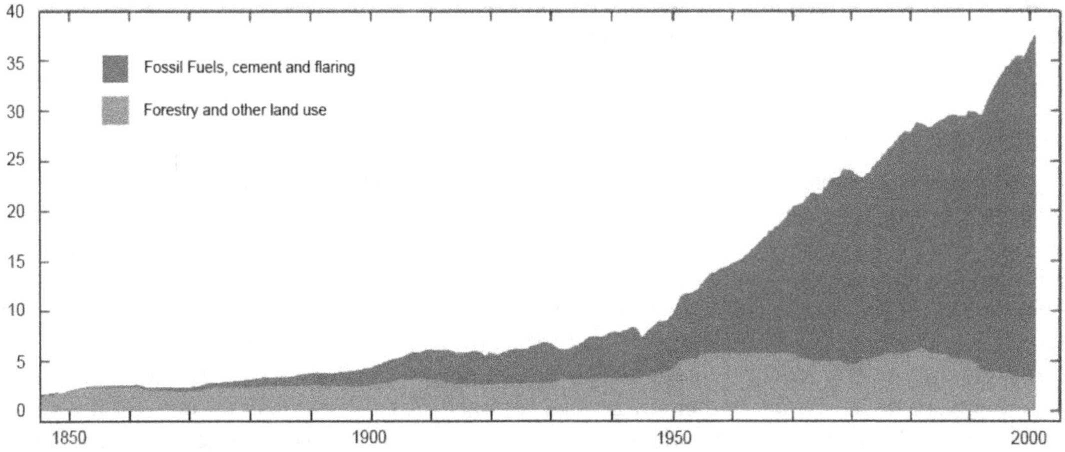

Note: This figure shows global emissions of carbon dioxide (CO_2) only and therefore omits the emissions of other important greenhouse gases like methane (CH_4) or nitrous oxide (N_2O). The IPCC notes that quantitative information on these other gases is limited for the period 1850-1970.

Source: IPCC (2014), Figure SPM.1 (d), www.ipcc.ch/pdf/assessment-report/ar5/syr/AR5_SYR_FINAL_SPM.pdf.

The particular case of fossil-fuel subsidies

There are several reasons why EHS are generally regarded as bad. Like most subsidies, EHS distort incentives to the extent that they alter relative prices for inputs or consumption goods. This in turn affects the decisions of producers and consumers, creating inefficiencies in the economy[3] and the use of resources. They are also costly in that they compete with other uses of public funds and deteriorate fiscal balances. What makes EHS different from other subsidies, however, is that they also cause environmental damage by definition. This report focusses on fossil-fuel subsidies since they constitute a notable subset of EHS owing to their prevalence and to the scale of their fiscal and environmental impacts. Previous estimates by the OECD and the IEA suggest that subsidies and other forms of support for fossil fuels exceed half a trillion USD worldwide annually (OECD, 2013b; IEA, 2014a), making them far from trivial.

Fossil-fuel subsidies are distortive

Changes in the price of fossil fuels relative to other goods and services can be expected to have large impacts on production and consumption decisions throughout the economy. Fossil fuels still are essential inputs to many economic activities, ranging from primary sectors such as agriculture and mining, to services like air transport and construction. They are also important for households who use them for heating and transport purposes. Furthermore, many countries rely extensively on fossil fuels for generating the electricity they need. By distorting costs and prices, fossil-fuel subsidies thus generate inefficiencies in the production and use of energy economy-wide. Only where subsidies serve to correct a pre-existing market failure can their use be potentially efficient from an economic perspective, such as subsidies for the provision of public goods like national defence and early-warning systems for natural disasters.

Particularly problematic for economic efficiency are subsidies that alter the rate of return on investment in selected assets or activities. Because they change the stream of income investors expect to receive for holding a particular asset, those subsidies influence investment choices and change the allocation of capital across sectors. In the case of certain fossil-fuel subsidies, there is therefore a risk that investors end up favouring sectors that produce fossil fuels or use them intensively, at the expense of cleaner forms of energy and other economic activities more generally. This problem is worse for long-lived capital infrastructure since the impact of investment decisions can in that case be felt for years or even decades.[4] For such assets, fossil-fuel subsidies that artificially increase returns on investment can lock in polluting technologies for years to come, thereby retarding the transition to a low-carbon economy. This also raises the chance that these assets become stranded in the face of environmental and regulatory changes (Ansar et al., 2013).

Fossil-fuel subsidies are costly

The Great Recession of 2008-09 and the rocky road to recovery left many countries in bad fiscal shape. Falling tax revenues due to slower economic activity, together with higher spending on stimulus packages, bailouts for the banking sector, and social transfers (including so-called automatic stabilisers) all resulted in record-high public deficits. In an environment characterised by low economic growth and low inflation, these deficits led to the accumulation of large volumes of public debt relative to countries' GDP (Figure 1.2). With many countries now in dire need of fiscal space, governments struggle to identify opportunities for cutting spending and increasing revenues in ways that do not damage the welfare of their citizens. A reform of fossil-fuel subsidies may form part of the solution, particularly in countries where they account for a relatively large share of the total government budget.

Figure 1.2. The crisis has left many countries in dire fiscal straits

General government debt as of 2014 (% GDP)

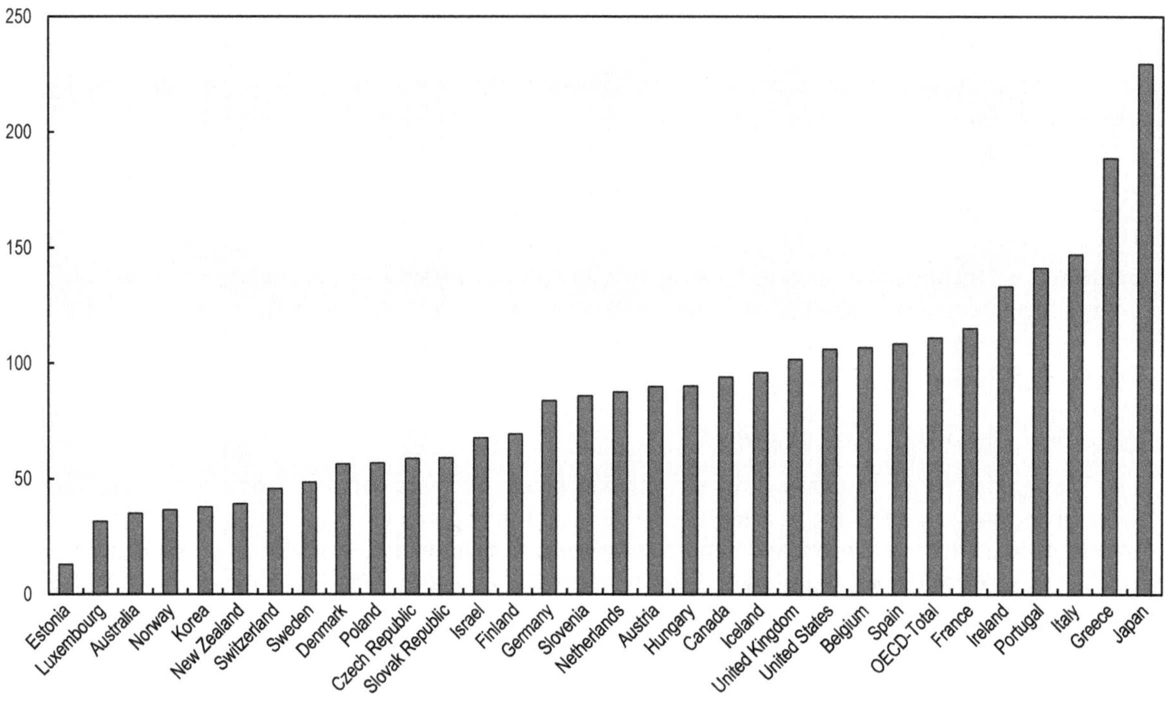

Source: OECD Economic Outlook No. 95, DOI: http://dx.doi.org/10.1787/data-00688-en.

In Indonesia, for instance, consumer subsidies for petroleum products and electricity (largely fossil-fuel-based) accounted for almost 20% of all central-government spending in 2011, an amount roughly equal to that spent on education, and one that was much higher than all spending on health and infrastructure combined (OECD, 2012a). As pressures mounted, the Indonesian Government subsequently managed to phase out entirely gasoline subsidies in its revised 2015 budget, leaving in place the smaller subsidies for LPG, diesel fuel, and kerosene and freeing up resources for more infrastructure investment (Sambijantoro, 2015). A comparable situation prevailed in India until the

Central Government started reducing consumer subsidies for diesel fuel in late 2012. The savings thus realised amounted to about INR 200 billion between the years 2012 and 2014. This represents roughly 10% of the revenues the country derives ever year from all its federal excise duties combined (Ministry of Finance of India, 2015). In the United States, the federal administration has repeatedly proposed that a number of tax preferences benefitting fossil-fuel producers be removed, with the proposed budget for FY2016 estimating the potential revenue gains at over USD 4 billion annually (OMB, 2015).

Fossil-fuel subsidies are environmentally harmful

Although most fossil-fuel subsidies are undoubtedly distortive and costly, the same could be said of many other types of subsidies, like home-ownership subsidies for instance. What really differentiates fossil-fuel subsidies from other types of subsidies are their environmental impacts. Combatting climate change will require large-scale reductions in GHG emissions through a shift toward low-carbon energy sources. Measures that encourage the production or use of carbon-intensive fuels clearly make that shift harder and more costly. This is especially so since many of the assets used in producing or combusting fossil fuels tend to have relatively long life spans, which creates a risk that capital infrastructure ends up locking in carbon-intensive technologies for years or decades to come.

GHG emissions are not, however, the only environmental externality associated with the extraction and consumption of fossil fuels. On the production side, mining activities can, for example, cause the land above to subside, with considerable impact on human activities and biodiversity.[5] While land subsidence is frequently associated with coal mining, the risks also exist for other fossil fuels as evidenced by heightened seismic activity around the Groningen gas field in the Netherlands (Dutch Safety Board, 2015). The contamination of ground and surface water can also occur in the event of an oil spill or where wastewater and residues from the extraction process are improperly disposed of. In addition, extraction techniques for unconventional sources of oil and natural gas (e.g. oil sands and shale gas) require considerable volumes of water, which aggravates the stress on water resources.

On the consumption side, the combustion of fossil fuels in power plants, in vehicles, and in buildings is directly responsible for the emission of numerous pollutants having local and often immediate impacts on the environment and on human health. In particular, ambient air pollution caused by particulate matter (PM), nitrous oxides (NO_X), volatile organic compounds (VOCs),[6] and sulphur dioxide (SO_2) increases mortality risks through a higher occurrence of respiratory diseases, cardio-vascular diseases, and cancers. This imposes very substantial costs on society as a whole. In the case of PM and ozone alone, a recent OECD study looking at ambient outdoor air pollution estimated the associated economic costs to have reached a total USD 3.6 trillion in 2010 in OECD countries, China, and India (OECD, 2014a). Besides impacts on human health, emissions of pollutants from fossil-fuel combustion also damage capital infrastructure (e.g. buildings) and in most cases impair crop yields through acid rain.[7] To the extent they encourage the use of fossil fuels, fossil-fuel subsidies contribute to these various externalities.

Some evidence on the benefits of reforming subsidies and other measures supporting fossil fuels

Analysis undertaken at the IEA and the OECD has until now largely focussed on the under-pricing of fuels in developing and emerging countries, for which data have been available for some time. Using a general-equilibrium model of the world economy, a recent study found that a co-ordinated multilateral removal of consumer subsidies in developing and emerging countries would reduce global GHG emissions by 3% by 2020 relative to a baseline scenario (Durand-Lasserve et al., 2015).[8] Most of that reduction would be driven by emission reductions in non-OECD countries since they are where the measured energy subsidies are concentrated. While aggregate real income would barely change at the global level following the removal of subsidies (+0.33% by 2020), this hides large disparities between importers of fossil energy that gain significantly (e.g. India and Indonesia) and exporters that lose slightly (e.g. Canada and the countries of the Commonwealth of Independent

States). Although the Middle East is a large energy exporter, the efficiency gains brought by the removal of subsidies seem sufficient to outweigh the negative impacts on the region's terms of trade, so that it records a net income gain overall.

The same study also employs a household survey to obtain micro evidence on the distributional impacts of the simulated reform in the Indonesian context. While the survey data indicate that consumer subsidies in Indonesia disproportionately benefitted rich households prior to their recent reform, their removal can, nonetheless, prove detrimental to the poorest segments of the population. Despite the fact that poor households often lack access to electricity and do not generally consume certain petroleum products (e.g. gasoline), a removal of energy subsidies can increase the price of many other goods these households consume, thereby affecting their purchasing power. For that reason, the study goes on to assess the distributional impacts of reform under three different scenarios, each of which involves a different kind of compensatory redistribution scheme, i.e. food subsidies, direct cash transfers to households, and labour-income subsidies. Simulation results suggest that cash transfers are the preferred option from an efficiency and equity standpoint. In particular, food subsidies would introduce new inefficiencies in the economy while wage subsidies would fail to reach the poorest since they do not benefit informal labour.

Other recent examples[9] of modelling-based analysis of the impacts of fossil-fuel subsidies and their reform include a study on Yemen that was conducted in 2011 by researchers at the International Food Policy Research Institute (Breisinger et al., 2011). Using a multi-sector model of the Yemenite economy, the study arrived at conclusions that are essentially similar to those of the OECD as regards the impacts of a reform of consumer subsidies. The results thus suggest that the gradual removal of petroleum subsidies in Yemen would increase economic growth relative to a baseline, but that minimising the impact on the poor would necessitate that compensatory measures be taken. In this case, compensation takes the form of direct cash transfers and infrastructure investment.[10] Although the study did not model environmental effects explicitly, it provides yet more evidence that fossil-fuel subsidies hamper economic growth through the impacts they have on fiscal resources and market signals.

In a different vein, in 2013 the United States Congress tasked the National Research Council to undertake a study of the effects of US federal tax policy on the country's GHG emissions (National Research Council, 2013). Among the several policies analysed by the research committee, the study assessed the impacts that certain tax expenditures benefitting energy producers have on domestic GHG emissions, including measures such as the percentage depletion allowance for natural-gas producers. This particular provision allows for the faster recovery of costs that are capitalised into fossil-fuel properties, thereby favouring investment in the exploration and development of US natural-gas resources. Results from a modelling exercise suggest that removing this tax concession would increase drilling costs and reduce incentives to explore and develop new gas supply. Given existing restrictions on the import (export) of natural gas into (from) the United States,[11] domestic natural-gas prices would increase following the reform since a modest increase in gas imports would not be sufficient to compensate for the lower domestic supply. This would in turn reduce natural-gas use in several sectors. Although the model's reference scenario points to a very modest reduction in CO_2 emissions (37 million tonnes over the model's time horizon, i.e. 25 years), this modest result is largely driven by the, now unlikely, substitution of coal for natural gas in the power sector. Considering the new carbon-emission standards for coal-fired power plants that have been proposed by the US federal administration, it is doubtful whether such substitution could occur on a significant scale in today's context. By effectively attaching a higher price to coal used in power generation, the new standards would make a shift from gas to coal unprofitable in most cases, thereby accentuating the projected decrease in CO_2 emissions should the percentage depletion allowance be reformed.

Some evidence on the production side also exists for the Russian Federation in the particular case of government support for the Yamal LNG and Prirazlomnoe upstream projects in the Arctic region (Lunden and Fjaertoft, 2014). In contrast to previous examples that focussed on the aggregate impacts of individual policies, this study adopts a systemic approach to evaluate the effects of government support and its removal on specific upstream projects. The analysis is here concerned with the combined effects of government funding of exploration and infrastructure development, tax concessions, and the government assumption of environmental risks among others. Results were derived using an *ad hoc* model of the Russian upstream sector and are indicative of the role that tax concessions and other government support play in rendering particular projects viable economically. For Yamal LNG, for instance, the analysis finds that government support did allow the project to move forward. Results are more ambiguous for the Prirazlomnoe project, where government support did not influence the go or no-go decision but was more akin to "a gift, in the amount of [USD] 16.5 billion in undiscounted terms, from the government to [Gazprom] rather than a step to fine-tune the taxation system" (Lunden and Fjaertoft, 2014). In the latter case, it would therefore seem that government support did not help increase hydrocarbon production, and instead ended up wasting public resources that could have been put to better uses.

1.2. A growing consensus for reform

The previous section has shown that there are several reasons why governments may want to consider reforming measures that support the production or use of fossil fuels. The issues raised by these measures are generally not new but the context within which they are adopted and reformed has changed. This might explain why recent years have witnessed an increasing number of international initiatives aiming to phase out or reform those fossil-fuel subsidies that are deemed harmful or "inefficient" by policy makers. This section provides a short overview of these initiatives, focussing on those most relevant to the work of the OECD and emphasising the value of international co-operation in the area of subsidy reform more generally.

The need for international co-operation

Starting in 2009, an international consensus has progressively emerged on the question of subsidies and other measures supporting fossil fuels. While there may be disagreements between certain groups of countries over issues of definitions and scope, governments are increasingly wary of the consequences that fossil-fuel subsidies can have for the environment and the global economy. That these concerns translate into international co-operation is no accident, however. Many of the questions raised by fossil-fuel subsidies and their removal are trans-boundary in nature and may thus require a co-ordinated response by governments.

GHG emissions are a global concern because such gases disperse globally, and stay in the atmosphere for decades or centuries, thereby eventually changing the climate. The IPCC's Fifth Assessment (IPCC, 2014) notes in that regard that:

> Effective mitigation will not be achieved if individual agents advance their own interests independently. Cooperative responses, including international cooperation, are therefore required to effectively mitigate GHG emissions and address other climate change issues.

Similar concerns arise in the case of other environmental externalities related to fossil fuels that cross borders, such as cases in which the PM and the SO_2 emitted by coal-fired power plants in one country affect air quality in another.

The trans-boundary impacts of subsidies and other measures supporting fossil fuels are not, however, confined to the environmental sphere. Like most commodities, fossil fuels are extensively traded internationally so that variations in supply or demand in one large country (or group of countries) can affect international prices, which will in turn affect supply and demand in other countries. Under certain conditions, it is thus possible that the removal of fossil-fuel subsidies in one country (or region) could reduce that country's (or region's) demand for fossil fuels enough that international prices would be lowered, thereby prompting more demand in other countries or regions. Past OECD analysis has shown this "leakage effect" to be plausible, though modelling results clearly indicate that the increase in fossil-fuel demand in other countries would not fully compensate the initial decrease in demand in countries that reform their fossil-fuel subsidies (Burniaux et al., 2011). Hence the removal of subsidies would still imply a net reduction in global demand for fossil fuels.

Because they lower the cost for consumers of energy derived from fossil fuels, subsidies and other measures that support the consumption of fossil fuels can also artificially improve the competitiveness of energy-intensive industries in countries that apply such policies. This echoes the earlier discussion of the distortions in costs and prices that subsidies cause, and where it was pointed out that such distortions can alter the allocation of productive resources (e.g. labour and capital) across the sectors of an economy. In the case of countries subsidising the use of fossil fuels in industrial processes, this may cause investment to be channelled toward energy-intensive industries (e.g. steelmaking and cement) while crowding out other economic activities. Here again, modelling analysis by the OECD shows this issue to be of particular concern for countries having relatively large consumer subsidies (Burniaux et al., 2011). For such countries, the analysis suggests that a reform of fossil-fuel subsidies could end up damaging the competitiveness of energy-intensive industries while, by the same token, improving the competitiveness of industries located in countries with no or smaller subsidies. The impacts of subsidies and their removal on industrial competitiveness and international trade constitute in that sense another important argument in favour of international co-operation.

Subsidies and other measures that push domestic fuel prices far below the prices prevailing in neighbouring countries can also encourage the smuggling of fuel across borders, sometimes in both directions.[12] Instances of fuel smuggling have been observed in various contexts, including at times between Singapore and Malaysia or between Brazil and Argentina (Kojima, 2013). In the former case, this has led the Singaporean Government to enact legislation requiring that all drivers leaving the country and entering Malaysia do so with their fuel tanks at least three-quarters full (Singapore Government, 2015). Smuggling is especially problematic for developing and emerging economies that seek to reform their own fuel subsidies but that are located in regions where illegal imports from neighbouring countries with heavy subsidies may end up cancelling partly the benefits of the reform. This has been the case in Colombia, where smuggling from the Bolivarian Republic of Venezuela led the two countries to co-operate at times to curb illegal trade in petroleum products (Kojima, 2013). Although political tensions between the two countries have since put a halt to bilateral arrangements, the example illustrates nevertheless that international co-operation can help make a reform of fossil-fuel subsidies more successful.

Building momentum internationally

International co-operation on the reform of fossil-fuel subsidies has been most visible in the context of the G-20, especially since leaders of its member economies committed in 2009 at the Pittsburgh summit to "rationalize and phase out over the medium term inefficient fossil fuel subsidies that encourage wasteful consumption" (G-20, 2009). Similar versions of this statement were reiterated at subsequent summits of the G-20, notably in Cannes, Los Cabos, Saint Petersburg, and more recently in Brisbane. Leaders of member economies of the Asia-Pacific Economic Cooperation (APEC) forum have made comparable announcements, starting with the Singapore declaration of November 2009, in which they committed to "rationalise and phase out over the medium-term fossil fuel subsidies while providing those in need with essential energy services" (APEC, 2009).

Following up on these commitments, members of APEC and the G-20 have over the past years engaged in annual rounds of self-reporting of their fossil-fuel subsidies, focussing on those that they deem inefficient. Lack of a shared definition and methodology has, however, made it difficult to reach a common understanding of the scope of "inefficient fossil fuel subsidies that encourage wasteful consumption", with individual country submissions varying greatly in depth and length. More recently, some of the G-20 members have agreed to subject themselves to reciprocal peer reviews of their fossil-fuel subsidies, with China and the United States volunteering in 2014 to be first. A similar peer-review process has commenced in APEC, starting with Peru in 2014 and continuing with New Zealand in 2015. Meanwhile, New Zealand has also co-founded and taken a leading role in the

Friends of Fossil Fuel Subsidy Reform (FFFSR) initiative, a group of like-minded, non-G-20 countries dedicated to advocating the reform of inefficient fossil-fuel subsidies on a global scale. Current members of this group include: Costa Rica, Denmark, Ethiopia, Finland, New Zealand, Norway, Sweden, and Switzerland.

The OECD has repeatedly contributed to these various initiatives by sharing its expertise and facilitating the exchange of relevant information among its member countries and other interested parties. As early as June 2009, members of the Organisation were already calling for "domestic policy reform, with the aim of avoiding or removing environmentally harmful policies that might thwart green growth, *such as subsidies: to fossil fuel consumption or production* that increase greenhouse gas emissions; that promote the unsustainable use of other scarce natural resources; or which contribute to negative environmental outcomes" (OECD, 2009; own emphasis). In addition to being an instrument of international co-operation, the OECD has also been a major provider of data on measures supporting fossil fuels ever since it published its first *Inventory of Estimated Budgetary Support and Tax Expenditures for Fossil Fuels* in 2012 (OECD, 2012b). The previous section has shown that this work has gone hand in hand with modelling analysis looking at the climate and economic impacts of simulated subsidy reforms.

In the wake of the Rio+20 United Nations Conference on Sustainable Development held in June 2012, participating countries have agreed to start a process for developing a set of Sustainable Development Goals (SDGs) that would draw in part on the already existing Millennium Development Goals. An *ad hoc* body, the Open Working Group, was subsequently established in January 2013 to oversee that process and work toward a set of proposed SDGs. The group has since issued a proposal in which Goal 12.c advocates the reform of "inefficient fossil fuel subsidies that encourage wasteful consumption […], including by restructuring taxation" while "taking fully into account the specific needs and conditions of developing countries and minimizing the possible adverse impacts on their development in a manner that protects the poor and the affected communities" (United Nations, 2014).

At the regional level, the European Commission's European Resource Efficiency Platform (EREP) has been tasked with providing high-level guidance on how to achieve the transition to a more resource-efficient European economy. Its 2012 policy manifesto already stressed the need to "[abolish] environmentally harmful subsidies and tax-breaks that waste public money on obsolete practices", and this advice was further reiterated in the first set of policy recommendations the Platform adopted in 2013, which states that: "The EU and Member States should as a matter of urgency phase out environmentally harmful subsidies (with the OECD definition in mind), with special emphasis on subsidies to fossil fuels and the use of water in agriculture, energy and industry. This should also cover fiscal advantages as well as distortionary pricing schemes" (European Commission, 2014a). A few regional development banks have also at times taken steps to evaluate or reform fossil-fuel subsidies in the countries in which they operate. This is the case of the Asian Development Bank, which has in recent years provided technical assistance for monitoring and evaluating fossil-fuel subsidies in some of its member countries (ADB, 2011). The Inter-American Development Bank is similarly undertaking technical co-operation for measuring and analysing subsidies for the production or use of fossil fuels in Latin American countries and the Caribbean (IADB, 2013).

These various initiatives show there is considerable interest in reforming fossil-fuel subsidies internationally. Co-ordinated progress at the international level can be slow and difficult, however. Capitalising on the existing momentum, some countries have therefore taken on themselves to move forward and reform their subsidies unilaterally. Not all such efforts have been successful though, and this underscores the importance of political-economy considerations in building domestic coalitions for reform. Chapter 3 provides examples of successful reforms in a number of countries, drawing on recent experiences in OECD countries and a number of partner economies.

1.3. How the OECD Inventory helps fill crucial data gaps

From looking at symptoms to characterising the disease: Towards establishing a full policy diagnosis

Before the OECD started in 2010 to collect data on measures supporting fossil fuels in its member countries, the only estimates of fossil-fuel subsidies that were available widely were those the IEA has been producing for its annual *World Energy Outlook* (WEO) since 1999. Because the geographical scope of those estimates is large — as is the ground the WEO generally covers — the IEA estimates fossil-fuel subsidies using available information on observed energy prices. By comparing local fuel prices in different countries to a set of reference prices (either import-parity or export-parity prices), the IEA calculates a number of "price gaps" to estimate the extent to which fossil fuels are under-priced in various countries.[13] To the extent lower consumer prices reflect the prevalence of subsidies, price-gap estimates should convey useful information about the magnitude of these policies. The IEA estimates that the fossil-fuel subsidies thus calculated totalled USD 548 billion in 2013 (Figure 1.3). Globally, it has identified 40 countries that subsidise their consumption of fossil fuels, and which together account for over half of the world's energy consumption. Of these, ten countries account for almost three-quarters of the total consumer subsidies measured; five of them — all oil and natural-gas exporters — are in the Middle East or in North Africa.

Price gaps have often been used in various contexts to measure subsidies or support to particular products, sectors or industries, such as where domestic prices exceed international reference prices so that a benefit is conferred to domestic producers. In the case of agriculture, the OECD has used price gaps to estimate market-price support to producers since the 1980s as part of a broader exercise aiming to evaluate total support for the farming sector (OECD, 2014b). The IEA was already calculating indicators of market-price support for coal producers in 1987 using a similar approach (IEA, 1988), though the exercise was subsequently discontinued.

Figure 1.3. The IEA estimates of fossil-fuel subsidies by type of fuel

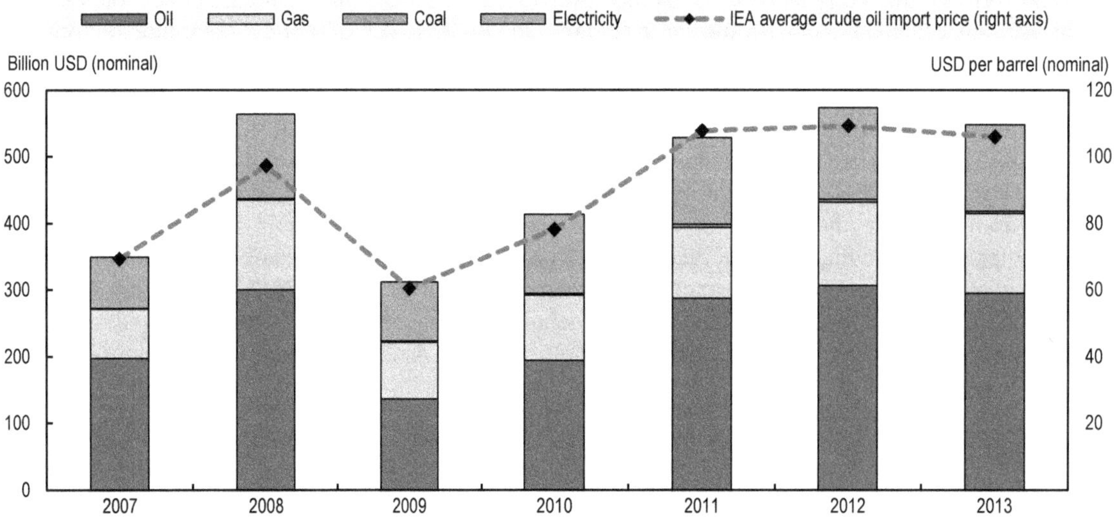

Source: IEA (2014a), DOI: http://dx.doi.org/10.1787/weo-2014-en.

Although it is undoubtedly valuable and helpful, the price-gap method does not capture certain forms of support for the production and consumption of fossil fuels. As argued by Koplow (2009), "relying solely on this metric would be a mistake" since it leaves out policies that do not lower consumer prices but that do have important fiscal and environmental impacts.[14] Examples of such policies include tax concessions, fuel vouchers and other payments made directly to low-income households, and many producer subsidies (IEA, 2014a). Measurement of producer support through price gaps is not a major feature since most fossil fuels are widely traded on world markets and often subject to low or zero import tariffs (see Tables A.2 and A.3 in the Annex). In this situation, producers only have a very limited ability to influence prices, except where they are large enough to affect world supply significantly (e.g. as with swing producers of oil), or where regulatory barriers and infrastructure bottlenecks insulate the domestic market from international price fluctuations (e.g. as with a lack of liquefaction terminals or pipelines for exporting natural gas). In other words, price gaps alone may not reveal all the producer support provided even as policies successfully increase domestic production of fossil fuels.

More generally, by focussing on the symptoms rather than on the disease, price-gap estimates do not provide information on the entire suite of policies and regulations that actually cause domestic fuel prices to fall below international reference prices. Establishing a full policy diagnosis necessitates that the price gap be attributed to specific programmes and measures and that stakeholders (e.g. beneficiaries) be clearly identified (Koplow, 2009; Kojima and Koplow, 2015). Failure to do so can hamper analysis of the full impacts that subsidies and other support measures have on the economy and the environment, and eventually make reforms more difficult by preventing a precise identification of potential winners and losers.

The need for an inventory: Addressing shortfalls in currently available data

The limitations described above in relation to the price-gap approach are especially problematic for OECD countries, where final prices for fuel generally exceed international reference prices due to the large range of indirect taxes that are often levied on the use of energy products there (Figure 1.4). Reasons for why these taxes exist are many, including considerations such as raising revenue and internalising the external costs associated with fuel combustion. Extensive information on the whole range of taxes levied on the use of energy and the corresponding tax rates can be found in *Taxing Energy Use 2015: OECD and Selected Partner Economies*, a companion OECD publication (OECD, 2015b) that also provides a series of graphical profiles of energy use and taxation — in both energy and CO_2 terms — for all OECD countries and a selection of key partners.[15] While this companion publication shows considerable variation in the extent to which different fuels are taxed across countries and sectors, the end result is generally such that domestic after-tax prices exceed international references prices in OECD countries and many partner economies. Because the reference prices used in calculating price gaps often do not comprise indirect taxes — other than value-added taxes (VAT) —, subsidies estimated using the price-gap approach are generally unable to account for support provided in OECD countries and in a number of partner economies.

As indicated earlier, price-gap estimates were the only set of data consistently available across different countries and years at the time G-20 leaders committed in 2009 to "rationalize and phase out over the medium term inefficient fossil fuel subsidies that encourage wasteful consumption." This generated an imbalance in country coverage since subsidies and other forms of support for fossil fuels escape measurement using price gaps in most high-income countries — a group of countries generally characterised by relatively higher taxes on energy use. In addition to creating a divide between high-income and middle-income economies in the G-20 and elsewhere, this lack of information erected additional barriers in the way of broader discussions of energy policy and reform. Transparency and information gathering form, indeed, step one in any policy assessment and reform process.

Figure 1.4. There is considerable variation in the extent to which energy is taxed across countries

Average effective tax rates on energy (left) and CO_2 from energy (right) in the OECD and in selected partner economies

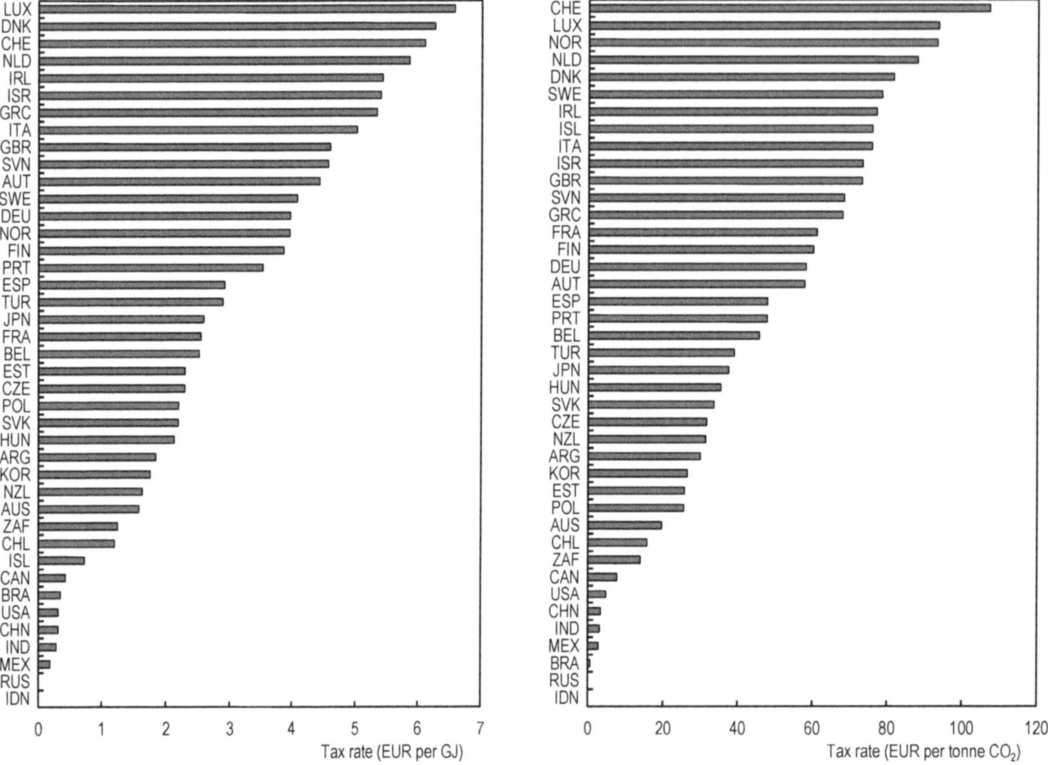

Notes: OECD calculations for the selected partner economies. Tax rates are as of 1 April 2012, except 1 July 2012 for Australia and Brazil and 4 April 2012 for South Africa. For that reason, the rates for Australia include the carbon tax that was subsequently repealed effective 1 July 2014. Energy-use data are for 2009 and are from the IEA. Rates for Canada, India, and the United States include federal taxes only.

Source: OECD (2015b), DOI: http://dx.doi.org/10.1787/9789264232334-en.

To address this problem, the OECD started in 2010 to collect information on all budgetary transfers and tax expenditures that encourage the production or consumption of fossil fuels in its member countries. These efforts soon led to the release of a first *Inventory of Estimated Budgetary Support and Tax Expenditures for Fossil Fuels* in 2012 (OECD, 2012b), and have since become a regular OECD exercise. By looking at the fiscal implications of individual support policies, the Inventory makes it possible to assess a whole range of government interventions at various points along the supply chain for fossil fuels, from the extraction stage to their combustion in vehicles or power plants. In particular, this approach accounts for support provided through the tax system, and more generally for various forms of support that do not push domestic fuel prices below international reference prices.

The Inventory thus assembled has fed into various publications and projects within the OECD and elsewhere. In addition to being an important input for recent OECD work on *Taxing Energy Use* (OECD, 2015b), its findings have frequently been used in the *Environmental Performance Reviews* and *Economic Surveys* of particular countries. Information from the Inventory will also contribute to a forthcoming OECD report that takes stock of the climate-change mitigation efforts undertaken to date in member countries, the European Union, and ten non-member economies (OECD, 2015c, forthcoming). Beyond the OECD, the European Commission has adopted the Inventory's method and framework to produce a follow-up study looking at budgetary support and tax expenditures for fossil fuels in those EU Member States that are not members of the OECD (IVM, 2013). This add-on was then followed a year later by a report seeking to enhance the comparability of data on budgetary support and tax expenditures for fossil fuels in the EU (European Commission, 2014b). The Inventory also formed the basis for some of the country submissions on fossil-fuel subsidies that were made in the context of APEC.

In that respect, the Inventory fulfils an important additional objective by promoting the transparency of public policies and government budgets, and eventually a greater accountability for how public resources are used. It can also be understood as a contribution toward the broader issue of how to make fiscal policy and tax systems "greener" or more environmentally friendly. In the same way that revenues from environmental taxes can be used to reduce other more distortive taxes (e.g. taxes on labour income), the reform of budgetary transfers and tax expenditures for fossil fuels could yield a so-called "double dividend" where revenue gains are significant.

For all its qualities, this Inventory should, nevertheless, be understood as a complement to the price-gap approach rather than a substitute. Both approaches serve distinct roles in discussions of fossil-fuel subsidies, and they should therefore not be opposed to one another. In particular, estimates derived using the price-gap method are particularly well suited for analytical work at the macroeconomic level, which facilitates subsequent analysis of the impacts of subsidies on international trade flows and global GHG emissions. They are also likely to be more accurate and comprehensive in certain countries that "lack the capability or will to provide accurate information on energy-related government activities" (Koplow, 2009).

Notes

1. Henceforth "China".
2. There are indications that stringent environmental policies — including measures for reducing GHG emissions — are not necessarily detrimental to short- or medium-term economic performance, whether measured in terms of productivity (Albrizio et al., 2014) or exports (Sauvage, 2014). See Koźluk and Zipperer (2013) for a survey of empirical findings on the topic. Arlinghaus (2015) and Flues and Lutz (2015) provide additional evidence on the impacts of carbon prices and energy taxes on firm competitiveness using a variety of indicators (e.g. output or employment).
3. These distortions also extend to foreign producers and consumers since virtually all economies are exposed to international trade. This forms the basis for the discipline of subsidies under the WTO's Agreement on Subsidies and Countervailing Measures (SCM).
4. The U.S. Internal Revenue Service sets, for example, the normal recovery period for pipelines used in carrying petroleum products and natural gas at between 15 and 22 years (IRS, 2014).
5. See Butt et al. (2013) for an overview of the various risks that fossil-fuel extraction poses for biodiversity.
6. NO_X and VOCs are also precursor gases of ground-level ozone (O_3), which causes a host of respiratory issues and affects the production of oxygen from leafy plants.
7. Some soils, such as those that are too alkaline, can benefit from acid rain. There is also some research suggesting that acid rain can reduce methane emissions from wetlands. See, for example, Gauci et al. (2008).
8. The OECD's ENV-Linkages model is a global recursive-dynamic, computable general-equilibrium (CGE) model, which simulates the interactions between firms and households across several sectors, regions, and years. The model's baseline scenario assumes no further policies than those already in place and its energy projections are calibrated on the IEA's

World Energy Model. See Chateau et al. (2014) for a presentation of the ENV-Linkages model.

9. This sub-section does not seek to provide a comprehensive review of the literature analysing the impacts of fossil-fuel subsidies and their reform. The intention here is rather to provide a few concrete examples of studies that do such analysis in order to illustrate the earlier discussion of the reasons why fossil-fuel subsidies are often regarded as harmful for society. See Ellis (2010) for a review of modelling and empirical studies on the effects of fossil-fuel subsidy reform that were undertaken between the early 1990s and 2009.

10. Because the study was conducted in 2010-11, some caution should be exercised when interpreting its results as a rebellion has since erupted in Yemen, casting doubt on the political feasibility of subsidy reform there.

11. In its current amended version, the Natural Gas Act of 1938 still requires that any company importing (exporting) natural gas into (from) the United States obtain prior authorisation from the US Department of Energy, irrespective of whether the gas is traded in gaseous or liquid form. Authorisation is, however, granted automatically to those selling natural gas to countries that have signed a free-trade agreement with the United States (IEA, 2014b).

12. Imports from neighbouring countries that subsidise their fuel is called "fuel tourism" when the transporter is also the final consumer of the fuel. Fuel tourism reduces the tax revenues that would otherwise be earned by the country in which the "tourist" is normally resident.

13. See Chapter 9 of the 2014 edition of the WEO (IEA, 2014a) for a description of the price-gap approach to estimating fossil-fuel subsidies.

14. On that issue, see also Kojima and Koplow (2015).

15. Box 2.3 in Chapter 2 provides more information on this companion publication.

Chapter 2

The Inventory approach to estimating support for fossil fuels

This second chapter introduces readers to the new Inventory of support measures for fossil fuels that the OECD has made available on its website in the form of an online database. Section 2.1 briefly sketches the structure of the database and its coverage, including what the OECD considers to be "support". Section 2.2 explains how the OECD collected the primary data that were then processed and transformed before they were eventually assembled in the database. In particular, the section describes the conceptual framework that the OECD uses to organise the information collected. Last, section 2.3 delves into the caveats that apply to tax-expenditure estimates since these account for more than half of all the measures the database contains.

The statistical data for Israel are supplied by and under the responsibility of the relevant Israeli authorities. The use of such data by the OECD is without prejudice to the status of the Golan Heights, East Jerusalem and Israeli settlements in the West Bank under the terms of international law.

2.1. A tool for transparency: The OECD's online database of measures supporting fossil fuels

Using the online tool

The online *Inventory of Support Measures for Fossil Fuels* identifies, documents, and estimates almost 800 individual measures supporting the production or consumption of fossil fuels in OECD countries and selected partner economies. In most cases, the information has been collected and assembled by the OECD itself, and then verified in co-operation with the governments of the countries concerned.

Each of the countries covered by this Inventory corresponds to a separate dataset. Individual entries in those datasets correspond in turn to the particular policies or measures applied by a given country, providing for each of them annual estimates of their budgetary costs or revenue foregone and a detailed description. This description covers several relevant characteristics of the measure, including — where available — its history, its eligibility criteria and beneficiaries, its transfer mechanism, its formal incidence, the fuels it benefits, etc.

The database is available through the OECD's online statistics portal (DotStat), where users can select the particular dimensions they are interested in. Quantitative information on the amounts of support provided annually by the different policies inventoried is displayed for the period 2000-14, except where the data are not available or not applicable. All amounts are in nominal units of national currency, unless otherwise specified. Qualitative information on the characteristics of each individual policy or measure can be accessed by clicking on the corresponding information bubbles in blue. Doing so opens a metadata window on the right-hand side that displays the qualitative information assembled by the OECD.

A comprehensive concept of "support" as a starting point for subsequent analysis

The Inventory proceeds from the fundamental perspective that the identification of "subsidies" to any sector or industry requires first taking an inventory of the full set of measures that may qualify as support to that sector. For one, because of interactive effects among policies, it is difficult to determine *a priori* whether a particular support policy is inefficient, encourages wasteful consumption, or is environmentally harmful. Only with a full picture of the operating policies can various analytical tools be brought to bear on questions about the effects of those policies on human welfare and the environment. Hence, information precedes analysis.

The scope of what is considered "support" is therefore deliberately broad, and is broader than some conceptions of "subsidy". Essentially, it includes both direct budgetary transfers and tax expenditures that in some way provide a benefit or preference for fossil-fuel production or consumption relative to alternatives. This broader definition therefore encompasses policies that can induce changes in the relative prices of fossil fuels. However, although the present inventory covers measures that provide support (either absolute or relative) to fossil fuels, it does not attempt to assess the impact on prices or quantities of the measures considered, nor does it pass any judgment as to whether a given measure is justified or not. In that sense, the inventory casts a wide net that aligns well with its objective of promoting the transparency of public policies.

It is recognised that policies supporting fossil fuels have been put in place for various reasons, i.e. support measures may have a *raison d'être* of their own. A consequence of the Inventory's broad conception of support is that while a number of these measures may be inefficient or wasteful, others may not be. The report does not provide any analysis of the impacts of specific support measures, and so does not pass any judgement on which measures might be usefully kept in place and which ones a country might wish to consider for possible reform or removal. Its purpose is rather to provide comprehensive information about policies that confer some level of support, as a starting point for subsequent analysis looking at the objectives of particular measures, their impacts (economically, environmentally and socially), and possible reforms and alternatives.

Forty countries and 40 fuel types: What the database covers

In its current form, the database contains individual entries for support measures previously or presently operating in the OECD's 34 member countries and six partner economies: Brazil, the People's Republic China (henceforth "China"), India, Indonesia, the Russian Federation, and South Africa. In addition, support provided by sub-national governments (e.g. states, provinces or *Länder*) is included for the following federal countries: Australia, Canada, Germany, and the United States. Due to time and resource constraints, sub-national entries for the United States only cover 11 states at this stage, the selection of which was informed by their relative abundance of fossil-fuel resources. These 11 states are: Alaska, California, Colorado, Kentucky, Louisiana, North Dakota, Oklahoma, Pennsylvania, Texas, West Virginia, and Wyoming.

The range of fuels covered by this Inventory comprises both primary fossil-fuel commodities (e.g. crude oil, natural gas, coal, and peat) and secondary refined or processed products (e.g. diesel fuel, gasoline, kerosene, and coal briquettes). Primary fuels include in particular those fossil fuels that are extracted from unconventional sources, such as oil extracted from bituminous sands, shale-based natural gas, or coal-bed methane. Measures supporting the production or use of biofuels are not, however, included in the present inventory. Nor are measures supporting electricity, except where it can be shown that the electricity is almost exclusively derived from fossil fuels, with limited possibilities for cross-border power exchanges.[1] To help ensure consistency with other existing datasets, the database follows the classification of fuels described in the *Energy Statistics Manual* (IEA et al., 2004).[2]

To keep the scope of the exercise manageable, the Inventory does not cover measures supporting the production or use of energy-consuming capital, such as vehicles and other equipment and machinery powered using fossil fuels. Although incentives for accumulating energy-consuming capital are likely encouraging more use of fossil fuels than would otherwise be the case, they are, nevertheless, much less specific in their relation to these fuels than are measures targeting energy sources directly. Measures supporting the manufacture and acquisition of road vehicles, for instance, can be expected to affect fuel consumption but they only do so in an indirect fashion. In that sense, they may be better characterised as support for the automotive industry rather than support for fossil fuels.

Some other measures may be directed at fossil fuels but may do so in a way that encourages the uptake of relatively cleaner forms of energy or practices. This is the case, for example, where support measures encourage a shift from coal to natural gas in power generation, or where support incentivises the use of LPG or compressed natural gas in road vehicles. Although such measures may serve to reduce GHG emissions in the short-run, they could also contribute to delaying the transition to other forms of energy since they lower the costs of producing or burning fossil fuels compared with alternatives. While recognising the potential for short-run environmental benefits of these measures, this inventory reports them anyway since not doing so would necessitate that some set of criteria be developed for assessing their environmental effects and justifying on environmental grounds their non-inclusion in the present inventory. Crucially though, the Inventory is not concerned with the effects of particular policies as explained earlier, nor does it pass judgment as to whether a given measure is justified or not. The inventory is, in that sense, not the proper place to discuss the environmental merits of individual measures. Policies supporting the development and deployment of technologies for carbon capture and storage (CCS) are not, however, included in the present inventory (Box 2.1).

Box 2.1. The case of support for carbon capture and storage

Carbon capture and storage (CCS) refers to "a family of technologies and techniques that enable the capture of CO_2 from fuel combustion or industrial processes, the transport of CO_2 via ships or pipelines, and its storage underground, in depleted oil and gas fields and deep saline formations" (IEA, 2013). Although CCS is frequently associated with the use of fossil fuels in thermal power plants and industrial processes, policies supporting the development and deployment of CCS technologies are not considered support for fossil fuels in the present inventory, where they are instead treated as support for energy-consuming capital.

CCS technologies are also generally regarded as potential tools for climate-change mitigation. The IPCC's Fifth Assessment Report thus notes that many of its models "could not achieve atmospheric concentration levels of about 450 ppm CO_{2eq} by 2100 if additional mitigation is considerably delayed or under limited availability of key technologies, such as bioenergy, CCS, and their combination (BECCS)" (IPCC, 2014). This reflects the view that significantly reducing emissions from energy-intensive industries, such as steel and cement, may sometimes prove difficult without CCS. For coal- and gas-fired power plants, CCS offers a possibility for avoiding the stranding of assets through retrofitting. In order for fossil-fuel facilities to be equipped with CCS, however, the cost of the technology would need to fall and the costs of unabated fossil-fuel use rise further (e.g. through a price on carbon emissions). It is currently estimated that CCS technologies could end up increasing the costs of coal-fired power plants by 40% to 63% in the 2020s.

By the end of 2014, 13 large-scale projects for the capture of CO_2 were operating globally, and a further 13 were in an advanced planning stage (IEA, 2015a). These include the Boundary Dam coal-fired electricity plant in Saskatchewan, Canada, which captures more than one million tonnes of CO_2 per year (the equivalent of one-third the emissions of a 500-MW coal-fired power plant), and Australia's Otway Project, which has so far stored 65 000 tonnes of CO_2 with some financial support from the State of Victoria. CCS technologies can also be employed for capturing the CO_2 emissions from sources other than fossil fuels. This is the case of the Decatur CCS project in Illinois, United States, which is scheduled to begin operation in 2015 and will capture CO_2 from bioethanol production rather than from fossil fuels. Several pilot CCS projects at cement plants also capture CO_2 from limestone calcination.

Sources: IEA (2013, 2015a), IPCC (2014), Zero CO_2 database.

2.2. Methods and data sources

How the primary information is collected

Generally, the data in the Inventory have been obtained from government sources. Support measures were identified mainly through searches of official government documents and web sites. In some other cases, unpublished data were furnished directly by governments. If no data could be found, the OECD estimated the value of support where it deemed the necessary calculations feasible and plausible. The data presented are therefore as comprehensive as possible, but they are by no means exhaustive. There is, in particular, more information presented in the Inventory for those countries that have been relatively more transparent in their budget books. This does not necessarily mean that these countries have higher levels of support than other countries, but may reflect that they have been more transparent about the support that is provided.

The sources used for compiling information on individual support measures are mainly the annual budgets of countries (e.g. budget statements, public accounts or budget statistics), which sometimes contain an annex describing and estimating tax expenditures. This follows from the fact that policy makers often regard tax expenditures as potential substitutes for direct spending since they constitute another way of transferring public resources (OECD, 2010a). In some other cases, tax-expenditure reports are instead published as stand-alone documents on an annual or biennial basis. There are, however, a number of countries that do not make their tax-expenditure estimates public, a fact which further complicates the collection of information. Hence, a limiting factor in respect of tax expenditures relating to fossil fuels is the extent to which countries release such estimates already.

With a few exceptions, most of the countries covered by the Inventory report their budgetary transfers and tax expenditures on a regular basis one way or another. Countries do differ, nevertheless, in the depth and scope of their reporting (Box 2.2). As regards tax expenditures, most of the reports cover both corporate and personal income taxes. However, fewer cover VAT, and even fewer attempt to estimate tax expenditures in respect of excise taxes. Differences also arise in the level of aggregation at which outlays and tax expenditures are reported. In some cases, the information available makes it possible to clearly identify the amounts of support benefitting users or producers of fossil fuels, such as where transfers are reported on an industry or sector basis. In others, the raw figures are too aggregated so that a further step is required to apportion the total support to the different industries or sectors benefitting from the measure. This is, for example, the case where measures relate to final energy consumption in general or to a range of natural-resource production rather than specifically to the production of fossil fuels.

Box 2.2. Reporting budgetary transfers and tax expenditures for fossil fuels: Examples from selected countries

Practices differ among countries as regards the reporting of budgetary transfers and tax expenditures. There is, however, a noticeable trend toward better reporting over time as governments gradually improve the quality and scope of the information they choose to release. In Italy, the *Delega Fiscale* now provides a legal basis for the annual reporting of tax expenditures and corresponding estimates of revenue foregone. In China, the Ministry of Finance has recently put in place an online portal for accessing the country's annual national financial accounts. Those contain detailed data on individual budgetary programmes, including the amounts disbursed by local governments.

Germany stands as a rare example: the Federal Ministry of Finance produces every two years a subsidy report (*Subventionsbericht*) containing detailed information and estimates for both budgetary transfers and tax expenditures. The information thus assembled is presented by sector, which makes it easier to assess how different industries in Germany compare in terms of total public support, and whether it is provided in the form of direct budgetary assistance or tax concessions. Being a federal exercise, the report does not, however, address the question of support provided by sub-national jurisdictions (*Länder*).

The Office of Management and Budget (OMB), an executive body, is responsible for preparing the budget of the US Federal Government. As part of its mandate, the OMB has been producing every year detailed reports of US federal tax expenditures ever since the Congressional Budget Act of 1974 required those tax provisions to be listed in the federal budget. Annual estimates from the OMB are easily accessible online but do not cover indirect taxes levied on motor fuels, nor do they cover the many tax expenditures US states provide at the sub-national level. The latter can, however, be found in the tax-expenditure reports that most US states now produce. The Joint Committee on Taxation (JCT) of the US Congress, a legislative body, also prepares in parallel its own list of federal tax expenditures, which does not always overlap with the list prepared by the OMB. Focussing on the energy sector more specifically, the US Department of Energy's Energy Information Administration (EIA) has periodically documented and commented on the various budgetary transfers and tax expenditures that benefit the production or use of fossil fuels at the federal level. Meanwhile, the Government Accountability Office (GAO) and the Congressional Research Service (CRS) have both at times produced in-depth reviews of US federal tax expenditures, thereby subjecting these policies to a considerable degree of scrutiny.

Sources: Bundesministerium der Finanzen (2013), CRS (2012), EIA (2011), GAO (2005), IMF (2012), JCT (2014), Ministry of Finance of the People's Republic of China (n.d.), OECD (2010a), OMB (2015).

How the information is then organised and processed

Once primary information is collected, each measure is then assigned a specific tag or identifier. This tag is in turn associated with a number of dimensions that describe the relevant characteristics of a measure. One such dimension is whether a measure belongs to the Producer Support Estimate (PSE), the Consumer Support Estimate (CSE) or the General Services Support Estimate (GSSE). Under the OECD's PSE-CSE accounting framework,[3] measures that benefit individual producers are classified under the PSE while those that benefit individual consumers are classified under the CSE. Measures benefitting producers or consumers collectively are classified under the GSSE, as are measures that do not increase current production or consumption of fossil fuels but that may do so in the future. Examples of measures belonging to the GSSE would include public support for industry-specific infrastructure development, such as public support for the construction of coal or natural-gas terminals, and government funding for sector-wide R&D in relation to fossil-fuel exploration and transformation.

For the purpose of the Inventory, the consumption of fossil fuels is defined as the stage at which fuels are combusted, whether it occurs in motor vehicles, stationary engines, heating equipment or power plants. Production therefore encompasses the following activities along the supply chain: exploration and extraction (EXTRACT), bulk transportation and storage (TRANS), and refining and processing (REFIN). Hence, measures supporting the transport of coal by barge or rail or those supporting petroleum refineries would form part of the PSE (or the GSSE where applicable). Continuing further downstream, consumption is here understood as spanning the following: the use of fossil fuels in power and heat generation (GENER); their use in industrial processes and activities, outside of the energy sector (INDUS); and all other final uses of fossil fuels (END), whether in the transport sector, the residential sector, or primary industries outside of the energy sector (e.g. agriculture and forestry). Figure 2.1 summarises graphically how these various stages along the supply chain fit within the PSE-CSE framework described above.

Another dimension along which the database characterises measures is through their formal incidence. Unlike economic incidence, which is concerned with the final beneficiaries of a measure, formal or statutory incidence does not take into account supply and demand elasticities, and is therefore solely focussed on which aspect of production or consumption is formally targeted by the measure. Formal or statutory incidence can in that sense be understood as *de jure* incidence while economic incidence is better understood as *de facto* incidence. As with a measure's environmental effects, it is only through careful analysis that the economic incidence of a policy can be established. The Inventory therefore looks for now at formal incidence only. To that end, formal incidence is divided into a number of categories, each corresponding to a separate column in Table A.4 (in the Annex), depending on whether a measure relates to: output returns (the unit revenues received from sales); enterprise income (the overall income of producers); the costs of intermediate inputs, such as feedstock; the costs of value-adding production factors – labour, land (which includes access to sub-surface natural resources), capital, and new knowledge; or consumption in general. As displayed, the matrix presented in Table A.4 provides an organising framework for relating a measure's formal incidence to its transfer mechanism, i.e. the manner in which the transfer is created, whether that be through a direct budgetary transfer, a tax concession or a loan guarantee.

Figure 2.1. Adapting the PSE-CSE framework to fossil fuels

Indicator	UPSTREAM →					→ DOWNSTREAM
	EXTRACT	TRANS	REFIN	GENER	INDUS	END
PSE	X	X	X			
CSE				X	X	X
GSSE	X	X	X	X	X	X

This Inventory concentrates of necessity on budgetary transfers and tax expenditures relating to fossil fuels since data for other more complicated forms of support can be much harder to obtain, as with the assumption by the government of certain risks otherwise borne by the private sector.[4] Numerous other forms of support — notably support provided through risk transfers — are not yet quantified however. The data requirements for estimating the transfers associated with such measures are greater and the calculations more complex. This is particularly the case as regards preferential loans and loan guarantees, where estimating the direct cost to the government of the assistance conveyed would require that a present-value calculation be performed using carefully selected discount rates (Lucas, 2015, forthcoming). Regarding market price support for producers, the previous chapter has already indicated that applied import tariffs on the main fossil fuels were generally very low or inexistent, even when looking at most-favoured-nation (MFN) tariffs, which do not account for the myriad of preferential trade agreements that are currently in force (see Tables A.2 and A.3 in the Annex).[5] Market price support is therefore unlikely to be of serious concern for fossil fuels as maintaining artificially high domestic prices would imply a significant degree of protection against international competition.

Several of the measures contained in the Inventory benefit more than one type of fossil fuel. The main transformation of the data carried out by the OECD was thus to allocate support to particular fuels where official government sources do not provide such a breakdown. Following standard practice (see, for example, OECD, 2010b), transfers associated with policies benefitting more than one fuel or sector were allocated according to the relative value of production or consumption, or proportional to the energy equivalent, volume of production or consumption. It is recognised that the actual allocation of support across fuel types may, in practice, vary based on factors other than the volume or value of production or consumption, but this approach is adapted in the absence of more specific information. For these reasons, while the primary data come from government sources, the particular breakdowns shown in the database may not necessarily reflect the views of responsible governments. In very few cases mainly pertaining to exemptions from indirect taxes, the OECD estimated the value of the tax expenditures, based on the published rate of exemption and national or IEA data on the volume of fuel that was exempted. Where applicable, details of any data transformation or calculation made by the OECD are described online in the metadata for the measures concerned.

2.3. Understanding tax expenditures relating to fossil fuels

Governments often use tax expenditures to support particular activities or entities that they deem beneficial from a societal perspective. It is therefore not surprising that institutions such as the European Commission or the WTO consider tax concessions as amounting to some form of "subsidy" or "state aid".[6] Although tax expenditures are by no means the only mechanism through which governments support the production or use of fossil fuels, their interpretation is, nevertheless, subject to a number of specific caveats that should be borne in mind when going through the database. This is especially so as tax expenditures are most often estimated with reference to a benchmark tax level or system that is country-specific.

Types of tax expenditures relating to the use of fossil fuels

Many of the tax expenditures that the Inventory contains are targeted at the final consumption of fossil fuels. They are typically provided through lower rates, exemptions, or rebates with respect to the two main types of consumption taxes:

- value-added taxes (VAT), which are intended to be broad-based taxes on final consumption, representing a percentage of the value of the good or service sold; and
- excise taxes, which are levied on specific goods, and for which the value of the tax normally is unrelated to the value of the underlying good but rather to its weight, mass, or energy content.

These are the most visible forms of tax expenditures relating to fossil fuels, as they often have a direct effect on final prices and therefore consumption, though the associated price impacts are not always easy to measure. Difficulties arise in particular as to the benchmark rates of tax that are used by countries for estimating the revenue foregone due to these tax expenditures. Not only do these benchmark rates vary across countries, but they also often vary across sectors and types of fuel, which then affects the range of provisions that governments consider to be tax expenditures (Box 2.3).

Some tax expenditures are applied broadly through general exemptions or reductions in countries' VAT rates. Other tax expenditures are more targeted. In this area, three main categories of tax expenditures stand out: (i) those related to specific groups of consumers; (ii) those related to specific types of fuel; and (iii) those related to how the fuels are used.

Box 2.3. Taxing energy use in the OECD and beyond

Complementing the present work that identifies, documents, and estimates measures supporting fossil fuels, the OECD has also looked in detail at the way energy use is taxed across countries, sectors, and fuels in the context of its publication *Taxing Energy Use*. This was done by assembling information on the specific tax rates that are applied to various energy sources in different sectors and countries, and then combining this information with corresponding data from the IEA on the volumes of energy used. The exercise makes it possible to express rates of tax on energy in a comparable fashion — usually units of national currency or EUR per GJ and per tonne of CO_2 — which facilitates the understanding of the structure and level of energy taxes, including tax expenditures where they exist. In particular, the publication provides a set of graphical profiles that illustrate concisely the use and taxation of energy in different countries and its implications for the price signals sent in relation to energy and carbon content. Figure 2.2 shows one such graphical profile using the example of Sweden.

Figure 2.2. The taxation of energy use in Sweden on a carbon-emissions basis

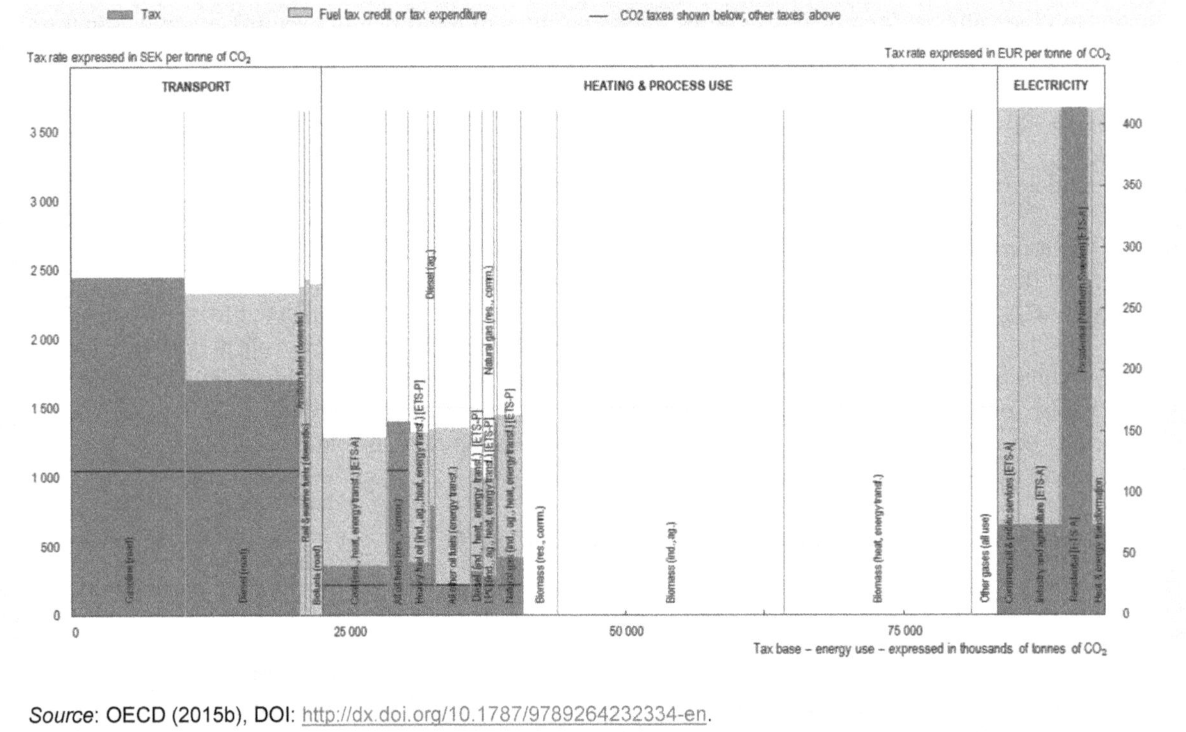

Source: OECD (2015b), DOI: http://dx.doi.org/10.1787/9789264232334-en.

In the first group, qualifying individuals or firms are taxed less heavily on their use of fossil fuels than are other users subject to the standard rate of tax. Often, beneficiaries include residents of particular regions that are deemed geographically or economically disadvantaged (e.g. France's overseas territories or certain areas of Italy). As with certain types of budgetary transfers, those concessions are generally intended to achieve social goals, such as supporting households' incomes. Other examples of tax expenditures in this first group include the exemptions from fuel tax that governments themselves (or diplomatic representations and international organisations such as the OECD) sometimes enjoy. This is the case in the United States, where state and local governments are generally exempt from excise tax for the fuel they purchase. In general, the tax concessions in this first group encourage higher consumption of the exempted fuels than would occur in the absence of the measures.

In the second group are tax expenditures that subject specific fossil fuels to reduced rates (or exemptions from tax altogether), even though these fuels are intended for the same end purpose as other fuels taxed at higher rates. A common example in the transportation fuel area is a lower tax rate (or exemption) on diesel fuel relative to gasoline (Harding, 2014).[7] Many countries also levy lower excise taxes on fuels deemed "cleaner" than gasoline or diesel fuel, such as compressed natural gas (CNG), liquefied petroleum gas (LPG), and biofuels, in an effort to encourage consumers to switch to those fuels. To the extent such differences in tax rates are considered to be tax expenditures by the countries concerned, they are included in the Inventory.

Finally, in the third group are tax expenditures occurring as a result of differences in rates based on how the fossil fuels are used. Provisions within this group are frequently found in primary industries such as farming, forestry, fisheries, and mining, where the use of diesel fuel often attracts exemptions or rebates from the excise tax normally levied on purchases of fuel. This is, for example, the case in Canada, France, Germany, Hungary, Israel, and Switzerland. Aviation fuels are a special case since they are generally sold free of tax when purchased for use in international flights.[8] Complications may also arise in relation to the taxation of fossil fuels used as inputs in industrial production processes (Box 2.4).

An important point to bear in mind when interpreting any tax expenditure relating to VAT or excise taxes on fuel is that, in most of the countries the Inventory covers, much of the fuel consumed — especially fuel used in motorised vehicles — is taxed to some degree. That which is not is generally sold at a price that is at least at world-market parity. The overall net effect of this taxation, even after the exemptions, reductions, and rebates, is still to provide some degree of disincentive to consume compared with a situation in which no taxes were applied, and hence no tax expenditures would be measured. Deviations from the standard tax rate nonetheless still distort relative prices *within* an economy, and may thus favour the consumption of certain fuels in preference to others. Excise-tax concessions and selective reductions in VAT rates for fossil fuels counteract, in that sense, "the intention of energy taxes to increase the relative end-user prices of energy (for environmental or for revenue-raising reasons)" (OECD, 2015b). This type of non-neutrality reported by governments constitutes "support" for purposes of the Inventory.

Box 2.4. Tax expenditures relating to fossil fuels used as inputs to production

A significant portion of fossil fuels (e.g. for heating in manufacturing plants, or as inputs to other uses) is consumed by manufacturers and service providers. Some tax expenditures are thus targeted at fossil-fuel products that form an input into production processes. With some types of taxation, such as with VAT, governments attempt to tax only the final consumption. In so doing, firms are effectively and necessarily exempted from the VAT they pay on inputs, through an input refunding system. Such measures are specifically designed not to discriminate among different production methods. As such, exempting energy, including fossil fuels, from VAT when it is only an input to production, can be consistent with the broader tax-policy aims of VAT.

Excise taxes, however, intentionally raise the price of the taxed item — e.g. because its use is deemed harmful to society, or because governments can raise revenues easily and relatively efficiently on its consumption. Given this intent, there is little rationale for exempting businesses that use these goods as inputs to production, as the goal is not to tax final consumption but the specific product or activity. In this case, a tax exemption may actually limit the effectiveness of the tax.

Industries engaged in the transformation of fossil fuels into more refined products or electricity are also often exempted from excise taxes on both the fuels they use as feedstock (i.e. intermediate inputs) and those they use as process energy (i.e. a value-adding factor). This is due to what is sometimes called the "manufacturer privilege" — a provision of the tax code which deems that all fossil fuels used in the production of final energy products (such as gasoline, coal briquettes or electricity) cannot be taxed. Yet those same fossil fuels, when used by other industries as part of their normal production processes, are often taxed. If the subsequent consumption of the energy products resulting from this type of energy-transformation process is subject to taxation (e.g. in the case of an electricity tax at the point of distribution), it might be logical to exempt the fuels used as feedstock in order to avoid double taxation. On the other hand, coverage of all fuel consumed as energy would require either taxation of the energy consumed in the transformation process (i.e. process energy) or, failing that, a grossing up of the tax on the energy outputs (e.g. the electricity) to account for the energy used in the production process. Given this, the Inventory generally includes tax concessions relating to fossil fuels used as process energy where such measures are considered to be tax expenditures by the countries concerned (e.g. in Germany).

The particular case of tax expenditures relating to the extraction of fossil fuels

Industries engaged in the extraction of hydrocarbons and mineral resources are unique from other businesses in that the key input to their production — the natural resource below ground — is commonly publicly owned.[9] Moreover, there is often significant uncertainty about a resource's exact extent and quality, and its value often depends significantly on the cost of production in the particular location. As with other depletable resources, the production of fossil fuels thus has the potential to generate above-normal profits in the form of a rent.[10] Therefore, in addition to levying the regular corporate income tax on profits earned in resource extraction, governments typically levy additional charges that may be seen as representing the "sale price" for the publicly owned resource. These charges may take various forms such as royalties, supplemental income taxes, resource taxes, and state participation through production-sharing contracts.

At the same time, many fossil-fuel-producing countries have corporate tax expenditures that are targeted at the extraction or production of fossil fuels, and their transformation into usable inputs for intermediate and final consumption. These are generally premised on concerns relating to risk and uncertainty, energy security, capital intensity, high upfront costs, and long project timelines, including extended pre-production periods. Such tax expenditures reduce the costs of extraction, and this in an environment in which jurisdictions often compete for attracting investment by mining companies so that resources do not remain untapped.

Tax expenditures in this area are commonly provided through the corporate income-tax system and may be targeted at fossil fuels or at resource extraction more generally. They include, among other features of the tax code, accelerated-depreciation allowances for capital expenditure, investment tax credits, additional deductions for exploration and development expenses, and preferential capital-gains treatment for particular fields. Tax expenditures on production can also take less visible forms such as the special treatment of income from state-owned enterprises, tax relief for income earned on industry sinking funds (e.g. for site remediation), tax exempt bonds, or the use of foreign tax credits for what may be considered royalty payments.

Tax-expenditure features may also be found in royalty systems, resource-rent taxes, and other specialised fiscal instruments applying to resource extraction. Such features must, however, be considered in the context of the particular fiscal system of which they form a part. This is especially so for measures relating to the tax treatment of capital expenses and financing costs incurred by fossil-fuel producers, which may or may not constitute tax expenditures depending on the broader nature of a country's fiscal regime applying to resource extraction. A provision allowing for the expensing (i.e. write-off) of successful exploration expenditures in the year in which they are incurred may, for example, be deemed normal practice (i.e. the benchmark) under a cash-flow tax system, such as Australia's Petroleum Resource Rent Tax. By contrast, although a similar kind of provision exists at the federal level in the United States for independent oil and gas producers, it is there considered a tax expenditure since the Federal Government taxes resource extraction using the common imputed-income approach, whereby expenditures incurred as a result of successful exploration efforts are capitalised and amortised over the useful life of the asset (e.g. the well).

Similar issues arise in resource royalty systems. Lower royalty rates on less productive or more costly fields may arguably be tax expenditures in that they represent a concession relative to standard rates. On the other hand, they may be rough ways of taking into account higher costs and lower margins in particular fiscal systems that otherwise would over-tax — and therefore potentially render uneconomic — marginal projects, i.e. projects that generate little or no economic rent. In a fiscal system designed for capturing resource rent, variations from the "benchmark" rate may be the norm. The approach of the Inventory is to include such reported royalty concessions (equivalent to tax expenditures), consistent with the purpose of highlighting cases where more favourable treatment is provided for one sector or group relative to the norm. It is intended to facilitate discussion about the purpose and impact of such concessions. As with relief from excise duties and carbon taxes, the support provided by particular tax or royalty concessions needs to be considered in the broader context of the fiscal system for resource extraction of which it forms a part (Box 2.5).

In general, the effect of tax expenditures supporting fossil-fuel production is to lower the cost of extraction and (since many are related to capital) provide an incentive for more investment, and potentially greater production, than would otherwise be the case. As noted in Chapter 1, this would generally be at the cost of reduced economic output elsewhere in the economy because of the diversion of investment. This can in turn affect both firm profitability and the price of fuels to be sold (depending on, among other things, the degree to which the price is set internationally). For firms with marginally profitable production, such schemes may not only have incremental effects on production, but can have a bearing on whether or not the firm continues producing at all. In other situations, such as where supply is constrained (by factors such as regulatory restrictions or limitations on labour or materials), tax benefits may simply increase firm profitability or contribute to inflation of input costs.

Box 2.5. Tax expenditures for resource extraction and the importance of the broader fiscal regime

The immediate write-off of expenditures of a capital nature — which include exploration and development costs — is normally considered to amount to some sort of preferential treatment under the tax systems of many countries. The reason is that in calculating taxable profits in most income-tax systems, capital expenses are amortised over the period to which they contribute to earnings. Allowing these types of expenditure to be written-off in full in the year in which they are incurred therefore provides companies with a benefit akin to a zero-interest loan from the government since it delays the collection of taxes. A present-value calculation would thus show a positive transfer from the government to the companies benefitting from such provisions.

However, when combined with a provision preventing the deduction of interest costs and other financing charges from taxable income, the immediate write-off of exploration and development expenses may not necessarily constitute preferential tax treatment (i.e. a deviation from "normal" taxation). This is because this particular tax configuration may approximate what is known as "cash-flow taxation". Under cash-flow taxation, it is a firm's cash flow rather than its true economic profit that forms the tax base so that "capital is costed by allowing an immediate write-off of investment expenditures at the time they are undertaken. No deductions for interest or depreciation are then permitted" (Boadway and Bruce, 1984). Cash-flow tax systems are theoretically equivalent to the more common imputed-income tax systems where the objective is to levy a neutral business tax. For that reason, measures such as the expensing of exploration and development costs may not necessarily be tax expenditures in countries that have adopted a cash-flow approach to taxing resource extraction.

Measurement and interpretation of tax expenditures

Unlike direct budgetary expenditures, where outlays can usually be readily measured, tax expenditures are estimates of revenue that is foregone due to a particular feature of the tax system that reduces or postpones tax relative to some benchmark tax system. This implies a number of important caveats concerning both the interpretation and comparability of the tax-expenditure estimates that governments produce. These caveats affect both: (i) what constitutes a tax expenditure; and (ii) how its size should be gauged. A number of these caveats are discussed in the remainder of this section.

Defining a benchmark in the broader context of countries' tax systems

A key challenge in determining or assessing tax expenditures is to identify the standard or benchmark tax regime against which the nature and extent of any concession is judged. The data on tax expenditures that are provided in the Inventory reflect estimates generated by national and sub-national governments themselves, and as such reflect the benchmark against which those governments chose to make these comparisons. Except in very few cases pertaining mainly to excise duties or VAT, the OECD did not select the tax benchmarks used in calculating the tax expenditures. Several approaches to deciding on the benchmark regime are possible, and these vary among countries:

- Many countries base their tax-expenditure estimates on a conceptual view about what constitutes "normal" taxation of income and consumption. Typically, the benchmark is defined to include structural features of the tax system, while special features intended to address objectives other than the basic function of the tax (e.g. raising revenues or internalising externalities) may be considered to be deviations from the benchmark. The line between what is structural and what is special is, however, often not a clear one.

- Some countries take a reference-law approach and identify only concessions that appear as such on the face of the law as tax expenditures. Under this approach, a tax credit would likely be identified as a tax expenditure, while differential tax rates on two products within a broader category might not be.
- A few countries restrict their tax-expenditure estimates to those tax reliefs (e.g. refundable income tax credits) that are clearly analogous to public expenditure.

Another approach is not to look at the current or normal tax regime but rather at an "optimal" tax regime. This is of particular relevance when investigating tax expenditures related to fossil fuels, given the presence of external costs or negative externalities — the cost imposed on others in society by a private action. When external costs are introduced, the issue of a baseline level against which to measure tax expenditures can change significantly. Curbing atmospheric emissions of harmful pollutants is one of the important reasons why countries implement environmentally related taxes, though other external costs, like traffic congestion and noise pollution, also sometimes motivate taxes (supplementing their motivation as a means to raise revenue for public purposes). Through excise taxes, countries can place a price on environmental damage, thereby encouraging a more socially optimal level of emissions. Under this approach, such taxes are levied along with the taxes normally needed for general revenue raising.

Although taxes are generally regarded as powerful tools for pricing external costs, the pursuit of optimal taxation[11] is complicated in practice. Quite apart from essentially normative issues such as determining revenue needs, countries would need extensive analytical work to determine optimal tax rates, which would vary significantly over time, and across users, locations, and types of fuel. For these reasons, external costs are not commonly considered in establishing tax-expenditure baselines. The IMF recently estimated nevertheless the level of taxation that would be required to internalise some of the external costs associated with the consumption of fossil fuels, focussing in this case on CO_2 emissions, local air pollution (SO_2, NO_X, and PM 2.5), and road-traffic-related externalities such as congestion and accidents (Parry et al., 2014). Using a number of assumptions (e.g. a social cost of carbon of USD 35 per tonne), the study found congestion, traffic-related accidents, and road wear and tear to account for the majority of the external costs it considers, representing more than 70% of all the shortfall in corrective taxes estimated by the IMF in the case of many EU countries (e.g. Belgium, France, Malta, and Sweden) and various other economies like Bhutan, Cape Verde, New Zealand, Syria, and Turkey. Excise taxes on fuel are, at best, an indirect way to reduce congestion though, which is a phenomenon that has more to do with the time of day when a vehicle is being driven, and where it is being driven, than with the act of consuming liquid fuel or electricity in a vehicle *per se*. Other instruments than fuel taxes may therefore be more appropriate for addressing certain external costs that bear only a loose relationship with the quantity of fuel consumed.

Whatever baseline is eventually chosen to measure tax expenditures, it is important to consider the overall taxation system. Since most countries do not have theoretically pure tax systems, there are sometimes tax features that may seem to support fossil fuels, but which are in fact mechanisms to compensate or correct for other features of the system. Similarly, a feature of the tax system that may be considered a tax expenditure in one country may not be a tax expenditure in another country, given differing overarching systems for taxing fossil fuels. Box 2.5 already mentioned this problem in the context of natural-resource extraction but it also presents itself for fuel use. The hypothecation or earmarking of tax revenues to fund specific public expenditures — making the tax a kind of user charge — is an issue that involves similar complexity, at least as long as earmarked revenues cover the envisaged expenditures. Other complications can arise where countries have allowed some reductions in a tax on fossil-fuel inputs to production processes, and the scale of these rebates reflects the degree of exposure of an industry to international competition or the deployment of other policy instruments to reduce emissions (as has occurred with some carbon taxes and emissions trading systems).

Calculating tax expenditures

Even where the baseline is clear, countries use different methods to arrive at their tax-expenditure estimates. The *revenue-foregone method*, the most straightforward, looks at the rate of the tax concession multiplied by the base or uptake, with no accounting for potential behavioural responses due to changes in the tax rates. For example, a reduced rate of EUR 0.25 per litre of diesel fuel used by taxis from a normal tax rate of EUR 0.45 per litre would yield annual tax expenditures of EUR 180 million if taxi drivers used 900 million litres of fuel a year. In practice, most countries rely largely on the revenue-foregone method to estimate their tax expenditures since the other methods require extra information and more complex calculations.[12] Because the Inventory uses the tax-expenditure estimates that countries themselves produce, the data reported therein usually follow the revenue-foregone method.

Measures that defer payment of tax without changing the ultimate nominal tax liability are another source of valuation differences across tax-expenditure accounts. A common example is accelerated depreciation allowances for capital investments. By allowing the cost of capital assets to be deducted more quickly than they would under the benchmark system, these provisions result in higher deductions and lower taxes in the early years in the life of a particular investment, but lower deductions and higher taxes in the later years of the investment. There are two main approaches to estimating the tax expenditure associated with such measures. The *nominal cash-flow approach* measures the extent to which taxes in a particular year are higher or lower as a result of the accelerated allowance than they would have been in its absence. This measure is normally negative in the early years of an investment (indicating a positive tax expenditure) and higher in the later years. In contrast, the *present-value approach* measures the discounted value of the time series of annual cash-flow tax expenditures, normally estimated from the time at which the asset is purchased. The two approaches both provide useful information, but they are quite distinct and not directly comparable.

While most governments typically use the cash-flow approach to estimate their tax expenditures in respect of tax deferrals, a few complement their estimates with illustrative calculations using alternative assumptions and methods. This is the case in the United States with the estimates the OMB reports every year, and which present the annual value of tax expenditures for tax deferrals on both a cash basis and a present-value basis (OMB, 2015). Whichever valuation approach is used, however, countries typically calculate the value of each tax expenditure on the assumption that all other provisions remain unchanged. Due to interactions and behavioural responses, the revenue impacts of eliminating multiple measures is not necessarily equal to the sum of the individual values. Caution is therefore required in adding together estimates for multiple measures.

International comparability

Tax-expenditure accounting was not designed with international comparability in mind. The estimates reported in the Inventory provide useful information about the relative treatment of different products within national tax systems, and the economic incentives created for actors within these systems. In the absence of a common benchmark, however, tax-expenditure estimates are not readily comparable across countries. Even where countries have adopted broadly the same methodological approach, the way in which they have implemented it in response to practical issues, such as how far a relief should be regarded as a structural part of the tax regime, may well differ (e.g. depreciation allowances used in calculating taxable profits). In general, a fundamental limitation on comparability is differences among countries in the definition of the benchmark tax system. For this reason, a simple cross-country comparison of tax expenditures can lead to a misleading picture of the relative tax treatment of fossil fuels. Figure 1.4 in Chapter 1 has already shown, for example, that average effective rates of tax on the use of energy differ widely across countries, which has a strong bearing on any tax expenditure relating to fuel consumption (OECD, 2015b).

With this in mind, tax-expenditure estimates must be used carefully. The fact that a particular country reports higher tax expenditures relating to fossil fuels does not always mean that this country effectively provides a higher level of support. The higher tax expenditures may also be due to factors such as:

- higher benchmark tax rates against which tax expenditures are measured;
- a stricter definition of the benchmark tax system that results in more features being singled out as tax expenditures; or
- a more complete set of tax-expenditure accounts.

Higher reported tax expenditures for some countries may therefore reflect higher levels of taxation or greater transparency in reporting rather than a higher level of "support".

Given differences among countries in levels of reporting with respect to tax expenditures, the OECD encourages all governments to be open and transparent in the reporting of tax-system features that may encourage the production or consumption of fossil fuels. Greater transparency will facilitate ongoing analysis and dialogue about how government policies, including those with respect to taxation, affect the production and use of fossil fuels. The European Commission (2014b) has already spearheaded efforts to express tax expenditures for fossil fuels on a common basis across EU Member States, and more work should be conducted in this area.

Notes

1. There are only a few such cases in the database.
2. Table A.1 in the Annex lists the different fossil fuels that the classification covers along with their respective codes as displayed in the online database.
3. The PSE-CSE accounting framework for measuring support to particular industries has long been used at the OECD to measure support to the agriculture sector and to the fisheries sector (since the mid-1980s and the late 1990s respectively). More information on that framework and its application for monitoring and evaluating agricultural policies can be found in OECD (2010b).
4. In practice, this implies that the inventory looks essentially for now at measures situated in the first two rows of Table A.4 in the Annex, with the addition of certain elements from rows three and four (e.g. royalty reductions and government buffer stocks).
5. Although Chile counts among the few countries that apply positive customs duties on their imports of fossil fuels, this reflects that country's reliance on a single MFN tariff (6%) applied uniformly on all imports rather than an explicit attempt to support Chilean fossil-fuel producers through higher prices. Chile being a party to numerous preferential trade agreements, the average import tariff it effectively applies on fossil fuels is likely close to zero.
6. Article 1 of the WTO's Agreement on Subsidies and Countervailing Measures (SCM) includes in its definition of a "subsidy" instances where "government revenue that is otherwise due is foregone or not collected." The European Commission's state-aid scorecard similarly takes into account measures such as a "tax credit and other tax measure, where the benefit is not dependent on having a tax liability", a "tax allowance, tax exemption, and rate relieve where the benefit is dependent on having a tax liability", and "deferred tax provisions (reserves, free or accelerated depreciation, etc.)." See for instance: http://ec.europa.eu/competition/state_aid/scoreboard/conceptual_remarks_en.html (accessed 8 April 2015).

7. The broader tax system must here be taken into account as some countries (e.g. New Zealand) have opted for distance-based road-user charges on diesel vehicles in lieu of an outright tax on purchases of diesel fuel. The choice of what constitutes "proper" or "normal" taxation for diesel fuel is not straightforward either. A recent OECD publication suggests, nevertheless, that the benchmark rate for diesel fuel ought to be at least equal to that for gasoline (Harding, 2014).

8. This is due to an international agreement dating from December 1944: the Convention on International Civil Aviation (also known as the "Chicago Convention"). This broad tax exemption was brought about to prevent distortions of aviation markets among countries, such as due to the double taxation of fuel, and to avoid inefficient tax avoidance behaviour, such as airlines shifting routes to reduce tax payments. Other arrangements generally exempt fuel used in international transport by rail and water as well. Mainly for that reason, the Inventory does not count as "support" fuel-tax exemptions for international aviation or international maritime transport. It does, however, include provisions exempting fuels used in domestic aviation and navigation.

9. Rules governing the ownership of underground resources in the United States differ from the rules applied in most other countries since private owners of non-federal US land also possess the corresponding mineral rights for sub-surface resources. This contrasts with other fossil-fuel-producing countries, where sub-surface resources generally belong to the public, irrespective of whether the land above is privately held.

10. Unlike manufacturing, many of the costs of production in natural-resource extraction depend on the location and geological characteristics of the resource being extracted. Given that market prices are volatile and determined by the marginal producer (usually the highest-cost producer supplying the market at any given time), the normal operation of the market can give rise to profits that are much larger (i.e. super- or above-normal) than those which would have been the minimum to justify investment in a particular well or mine.

11. That is, the level of taxation that accounts for all externalities, efficiency effects, the revenue-raising needs of the government, and the interaction of these effects on the overall economy.

12. The *revenue-gain method* estimates the increase in tax revenues that the government could expect if the tax expenditure were eliminated, thereby incorporating anticipated behavioural changes. Using the same example, the tax expenditure under this method would be the difference in tax rates — EUR 0.20 as before — multiplied by the expected use of diesel fuel by taxi drivers. Under this method, the use would be below 900 million litres since raising the tax rate would likely reduce the consumption of diesel fuel and increase that of gasoline. The *expenditure-equivalent method* estimates the level of funding that would be needed to meet the same outcome using a spending programme. In the previous example, it would estimate what level of direct payments would be needed to maintain the level of taxi drivers' income if the tax expenditure were eliminated.

Chapter 3

Tracking progress in reforming support for fossil fuels

This final chapter uses the data compiled for the 2015 edition of the OECD Inventory to derive a few results and indicators on the magnitude and nature of support for fossil fuels in OECD countries and the selected partner economies. The first section looks at broad trends in aggregate support and relates the observed evolution to recent policy changes and reforms. Section 3.2 looks at the characteristics of individual support measures to better understand the way support is provided to producers and consumers. Section 3.3 puts consumer support in perspective by assessing it in the broader context of countries' energy taxation. Finally, section 3.4 concludes by suggesting that further action be taken by policy makers to continue reforming measures that support fossil fuels.

The statistical data for Israel are supplied by and under the responsibility of the relevant Israeli authorities. The use of such data by the OECD is without prejudice to the status of the Golan Heights, East Jerusalem and Israeli settlements in the West Bank under the terms of international law.

3.1. A first glance at the data

Recent reform efforts are paying off

Taken together, the almost 800 measures contained in the Inventory had an overall value of USD 160-200 billion annually over the period 2010-14. This includes both support provided by OECD countries and that provided by a selection of partner economies (Brazil, the People's Republic of China,[1] India, Indonesia, the Russian Federation, and South Africa). Compared with the previous edition of the *Inventory* (OECD, 2013b), which focussed on OECD countries only, support now seems to follow a downward trend after having peaked twice in 2008 and 2011-12. Although the decline is more marked in OECD countries, a similar downward trend can also be observed in partner economies, where total support has been showing clear signs of recession since 2012 (Figure 3.1). In both cases, the decline in total support finds its origin in lower international oil prices but also in important policy changes, which signal an intention on the part of many governments to depart from earlier practices and move toward growth patterns that are more sustainable fiscally and environmentally.

A sizable portion of the decrease in support observed for OECD countries can be ascribed to Mexico, which eliminated the support it provides for the consumption of gasoline and diesel fuel through its IEPS (*Impuesto Especial sobre Producción y Servicios por Enajenación de Gasolinas y Diesel*), a floating excise tax. Variable rates of IEPS are set by the government on the basis of international oil prices for the country's two brands of gasoline, "Magna" and "Premium", and diesel fuel. When international oil prices are high, IEPS rates turn negative, which generates a tax expenditure. Conversely, lower international prices trigger an increase in the variable rates of IEPS, which reduces the tax expenditure or, as is currently the case, results in a positive tax. The Federal Government has over the years steadily increased retail prices on a monthly basis in order to reduce the support conferred to consumers (Figure 3.2). Together with the lower international prices, these efforts have contributed to reducing total consumer support in Mexico from MXN 244 billion (USD 18.5 billion) in 2012 to MXN 34 billion (USD 2.5 billion) in 2014. Since late 2014, rates of IEPS have been positive and it is expected that these will generate revenues of around 1% of GDP in 2015.

In the case of partner economies, most of the decline observed between the years 2012 and 2014 has to do with India's decisive efforts to rein in spending on consumer subsidies for diesel fuel. Starting in late 2012, the federal government thus decided to periodically increase retail prices by small amounts (about INR 0.50 a month, corresponding to USD 0.008), which eventually led to the termination of the subsidies for diesel fuel in September 2014. The reform has had a large impact on public finances, with total consumer support for petroleum products falling from about INR 970 billion (USD 18 billion) in 2012 to INR 610 billion (USD 10 billion) in 2014. Large subsidies remain for kerosene and LPG but the move represents nonetheless a major step in the right direction.

Mexico and India are not isolated cases, however. In the first quarter of 2015, the Central Government of Indonesia took decisive action in its revised budget for the year and scrapped all of its gasoline subsidies, while also capping the subsidies it provides for diesel fuel at IDR 1 000 per litre (about USD 0.08 per litre). This unprecedented move will reduce the total cost of Indonesia's consumer subsidies for petroleum fuels from IDR 247 trillion in 2014 to IDR 65 trillion in 2015, thus approaching a USD 14 billion decrease in a single year.

Figure 3.1. Support overall remains high at USD 160 billion despite signs of decline

Total support for fossil fuels in OECD countries (left) and selected partner economies (right) by year and type of fuel (Millions of current USD)

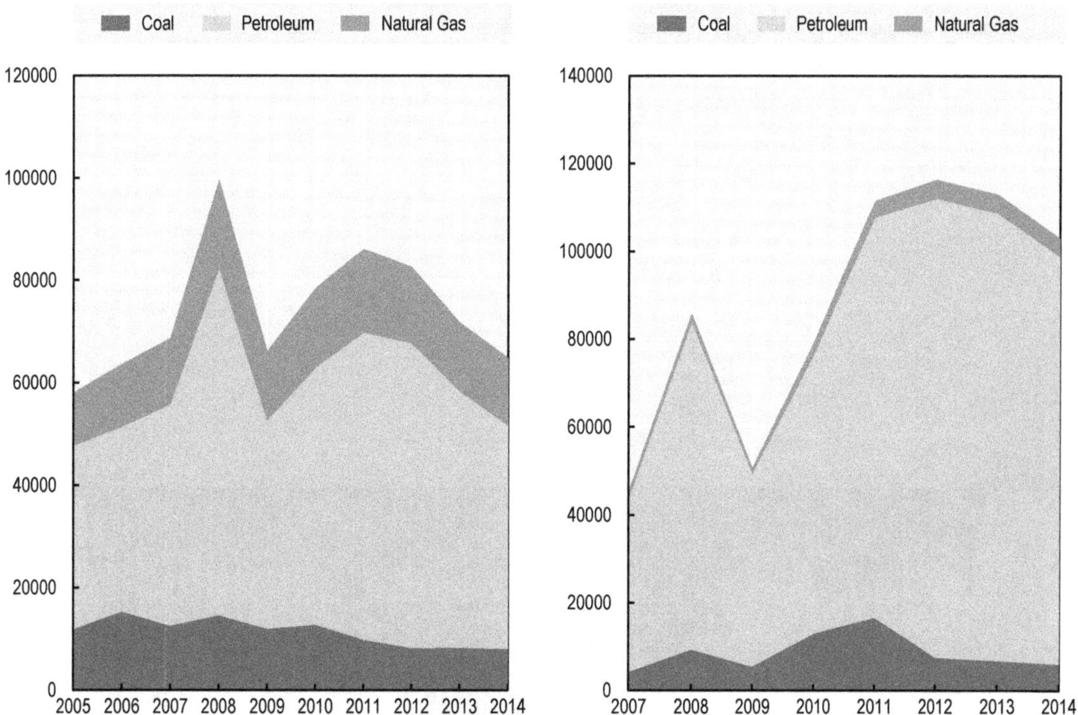

Notes: The above charts are based on an arithmetic sum of the individual support measures identified in the *Inventory*. Along with direct budgetary support, it includes the value of tax relief measured under each jurisdiction's benchmark tax treatment. The estimates do not take into account interactions that may occur if multiple measures were to be removed at the same time. Because they focus on budgetary costs and revenue foregone, the estimates for partner economies do not reflect the totality of support provided by means of artificially lower domestic prices. Particular caution should therefore be exercised when comparing these estimates to those reported by the IEA (2014a) for these countries.

At a lower scale, progress is also visible in a number of OECD countries. In January 2013 the Netherlands phased out the excise-tax reduction it had previously been applying to diesel fuel used for non-transport purposes (e.g. in farming activities or for heating) on the grounds that the concession was environmentally harmful and costly to monitor. Austria and the Slovak Republic took similar steps in 2013 and 2011 respectively. Canada has in recent years reformed federal provisions relating to the treatment of certain capital expenses for oil sands and coal mining in order to improve the neutrality of the country's corporate tax system. Germany has continued reducing the large budgetary transfers it provides every year to hard-coal mines located in North Rhine-Westphalia, bringing payments to EUR 1.5 billion in 2014, down from about EUR 4.8 billion in 1998. The country plans to phase out these transfers entirely by 2018. France took important steps in 2014 to gradually remove the exemption from excise tax it applied to natural gas consumed by households. With the phased introduction in 2014 of a carbon component in excise taxes (known as the Climate Energy Contribution, or *Contribution Climat Énergie*), this tax expenditure is expected to terminate as rates of excise on purchases of natural gas start increasing in line with a set price for carbon.

Figure 3.2. Mexico eliminated the support it provided for the consumption of gasoline and diesel fuel through its floating excise tax

The evolution of IEPS rates in Mexico over 2009-15
(MXN per litre shown as bars; USD per litre shown as lines)

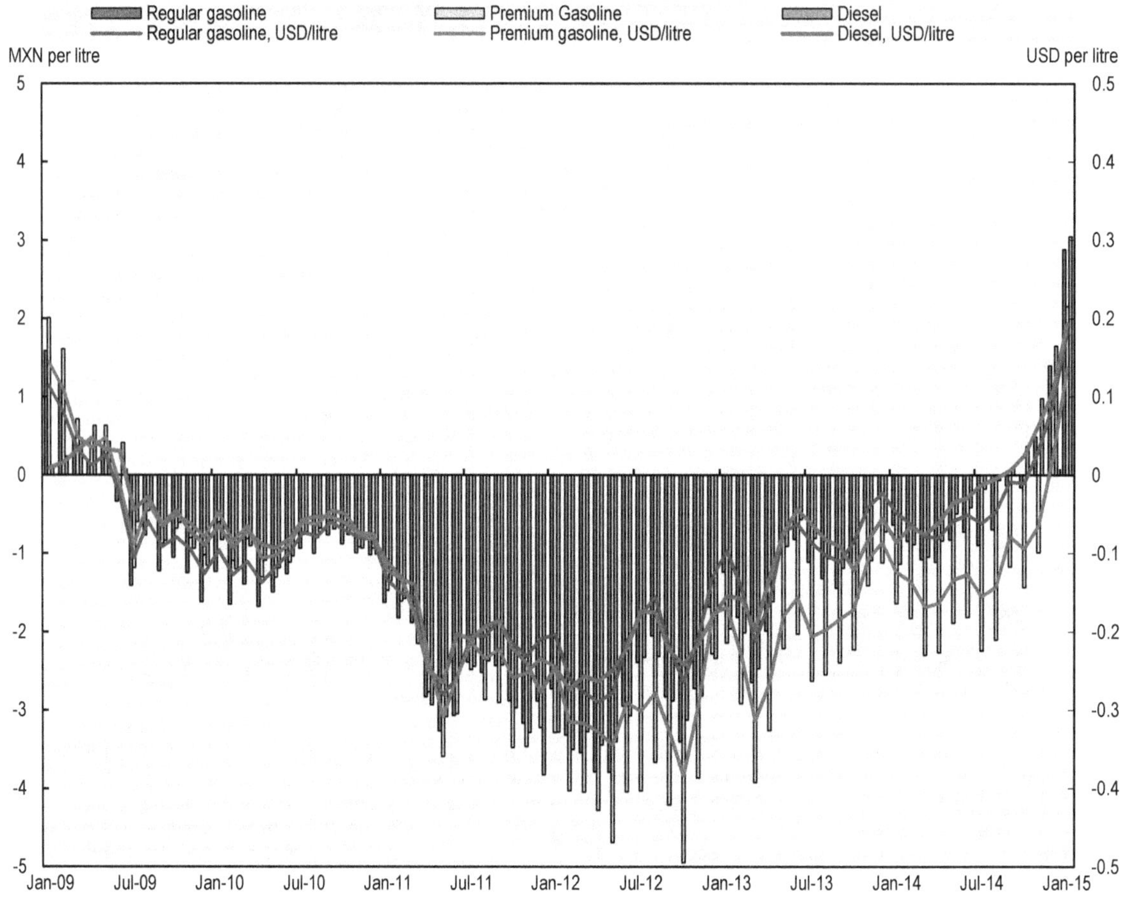

Source: Secretaría de Hacienda y Crédito Público, Federal Government of Mexico, sie.energia.gob.mx/bdiController.do?action=cuadro&cvecua=PMXE2C18E.

Support for the consumption of petroleum products still accounts for the bulk of total support

Whether one looks at OECD countries or partner economies, crude oil and petroleum products clearly attract most support, accounting for more than four-fifths of the total amount (82%) over the period 2012-14. By comparison support for coal and natural gas seems much more modest, representing around 8% and 10% of all support respectively. In part, this reflects the large share of petroleum products in countries' total primary energy supply, where fuels such as gasoline, diesel, and fuel oil dominate the transport sector and parts of the residential and commercial sectors. Fuels used in transport are also more taxed on average than other energy sources (OECD, 2015b), which can result in comparatively larger tax expenditures where tax concessions for such fuels exist.

Concomitant with a high share of total support going to petroleum products, the data also point to an overwhelming predominance of consumer support (more than 80%).[2] While this is hardly surprising for those emerging economies that are characterised by very large consumer subsidies, the situation needs more explaining for OECD countries. There, the prevalence of consumer support owes a lot to the fact that many large OECD economies do not extract fossil fuels on a significant scale. This is, for example, the case of France, Italy, and Sweden, where fossil-fuel extraction is very modest and production mainly occurs in the refining and processing sector. By contrast, focussing on countries that extract significant quantities of fossil fuels shows producer support to weigh more than what the overall results suggest. Producer support (i.e. PSE) as a share of total support thus exceeded 35% on average in Canada (38%), Germany (43%), the Russian Federation (78%), and the United States (42%) over the period 2012-14.

3.2. Anatomy of a support measure

How support is generally provided

Looking at individual measures and their characteristics rather than at the amounts of support they confer changes the picture somewhat, with consumer measures representing about half of all the measures the Inventory contains, whereas producer measures and GSSE measures account for 37% and 13% respectively. This means that a consumer measure generates on average more support (in absolute terms) than a producer measure or a GSSE measure. The relatively high tax benchmarks used in calculating tax expenditures for motor fuels may explain part of that result, as may the very large consumer subsidies observed in a number of partner economies.

In terms of formal or statutory incidence,[3] apart from consumption (which logically accounts for half of all measures, since consumption is the only incidence category for CSE measures), the results indicate that land & natural resources and capital represent 18% and 11% of all measures respectively, followed by knowledge creation (6%), the cost of intermediate inputs (5%), enterprise income (3%), output returns (3%), and labour (3%). This is hardly surprising given that resource extraction and energy transformation tend to be relatively capital-intensive activities. Adding in information on the stage of the supply chain at which policies intervene (see Figure 2.1) shows producer measures to revolve mostly around the extraction stage (42% of all measures), with bulk transportation and storage (4%) and refining and processing (4%) making only a small contribution to the total number of measures.

The Inventory shows a certain degree of policy inertia

The wealth of information contained in the Inventory reveals a few trends and commonalities on measures supporting fossil fuels in OECD countries and the selected partner economies. For example, most measures (about two-thirds of them) seem to have been introduced prior to 2000. This indicates that these policies were in many cases introduced in a very different context than today's. For some, they may have been adopted at a time when climate change was not deemed a concern among policy makers. The economic and political context might have been different too, e.g. as with higher economic growth or higher price inflation. Several federal measures in the United States were, for example, introduced between the 1970s and the 1980s,[4] a period characterised by widespread concerns relating to energy security in the aftermath of the oil crises of the 1970s. It is interesting to note also that some producer measures were put in place precisely when international oil prices collapsed in 1986, so that these measures may have at the time constituted attempts to shore up domestic production capacity.

What this discussion suggests in general is that there might be a need for countries to reassess the relevance of some of their support measures in today's context. Around 60% of all measures are tax expenditures, some of which are long-standing tax provisions that are rarely questioned in the domestic context (e.g. France's VAT and excise-tax reductions for gasoline sold in Corsica). Others are short-lived initiatives adopted in response to the circumstances of the time (e.g. Alberta's 2009-10 Energy Industry Drilling Stimulus). Either way, policy makers may wish to engage in periodic reviews of their countries' support measures as changing circumstances can render certain provisions obsolete or not suited to current challenges.

3.3. Consumer support for fossil fuels in the broader context of energy taxation

As Chapter 2 pointed out, tax-expenditure estimates are subject to a number of built-in assumptions and caveats that have a bearing on the interpretation of support amounts. Although the Inventory contains many more policies than just tax expenditures, the latter's prevalence is enough to make direct international comparisons difficult, and this imposes strong limitations on the kind of analysis that can be undertaken with the database. A crucial aspect concerns differences in rates of tax that exist across countries since higher rates increase tax expenditures other things equal. Another relates to the scope of what countries consider to be tax expenditures. Together with the size of economies (e.g. as measured using countries' GDP), one could expect those factors to influence the total amounts of support different countries provide.

To account for this possibility, the analysis expresses total consumer support (i.e. total CSE by country) relative to the energy component of the revenues countries derive from environmentally related taxes.[5] Using those revenues as a scaling factor should account for both the size of countries (larger countries raise more revenues all other things equal) and countries' general attitude toward energy taxation (higher rates generally mean higher revenues). Further adjustments are then made to improve comparability, such as removing tax expenditures relating to the lower taxation of diesel fuel for road use relative to gasoline, where such measures are considered tax expenditures. Not doing so would exaggerate the importance of consumer support in countries that treat this tax differential as a tax expenditure (Denmark, Finland, Norway, and Sweden), thereby penalising transparency in tax-expenditure reporting.[6] Figure 3.3 shows the numbers thus obtained.

Figure 3.3. Total consumer support (CSE) expressed as a share of the energy component of environmentally related tax revenues

Average for 2010-12

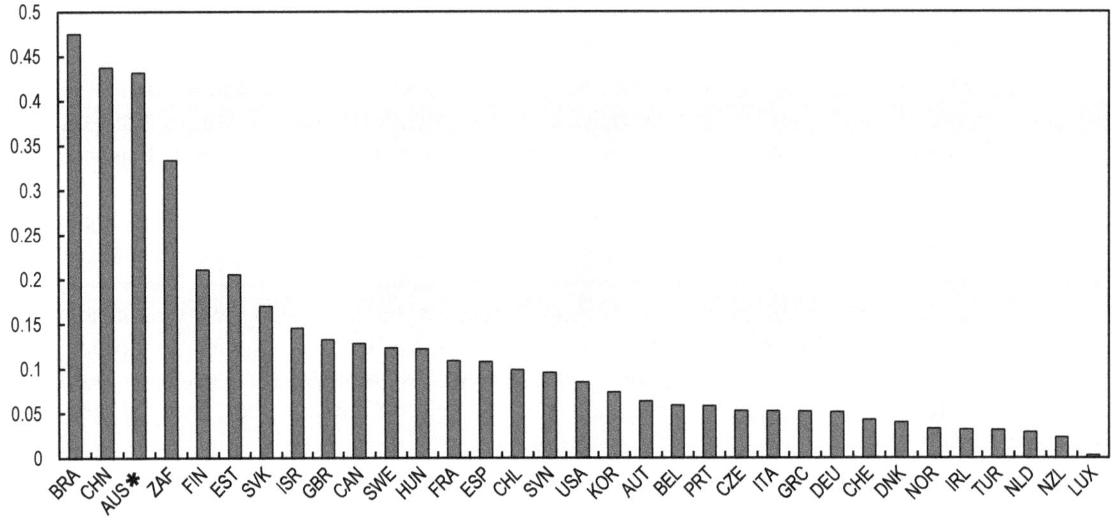

Notes: *The data for Australia include the country's large Fuel Tax Credits, which alone explain the relatively high ratio observed for that particular country. This measure serves to rebate some of the excise taxes that businesses pay on their purchases of fuel there. Data for Brazil and Greece are for the period 2010-11 only.

Tax rates would appear to be just one of many factors behind consumer support expressed in relative terms. Unsurprisingly, the data indicate that consumer support relative to environmentally related tax revenues tends to be higher in partner economies than in OECD countries. This reflects in part a lesser reliance on environmental taxation (and taxation in general) in the former group, along with higher consumer support there more generally. Less obvious are the relatively large ratios observed for some OECD countries having higher rates of energy taxation. This is especially so in view of the weak correlation that exists between total consumer support as a percentage of GDP and the average effective rates of tax on energy use calculated by the OECD[7] (Figure 3.4), which suggests that tax rates are not the main determinant of consumer support expressed in relative terms. What this result might indicate though is a higher reliance in these countries on tax expenditures for targeted fuel usages. As explained before, caution is, however, required in interpreting the ratios in Figure 3.3 since differences remain in how countries define their tax expenditures, and this even though adjustments were made to improve comparability. These problems are considerably more severe for producer measures. No attempt was therefore made to undertake a comparable exercise for producer support (PSE).

Figure 3.4. Tax rates are not the main determinant of consumer support

Total consumer support (CSE) as a percentage of GDP and average effective rates of tax on energy use (2012)

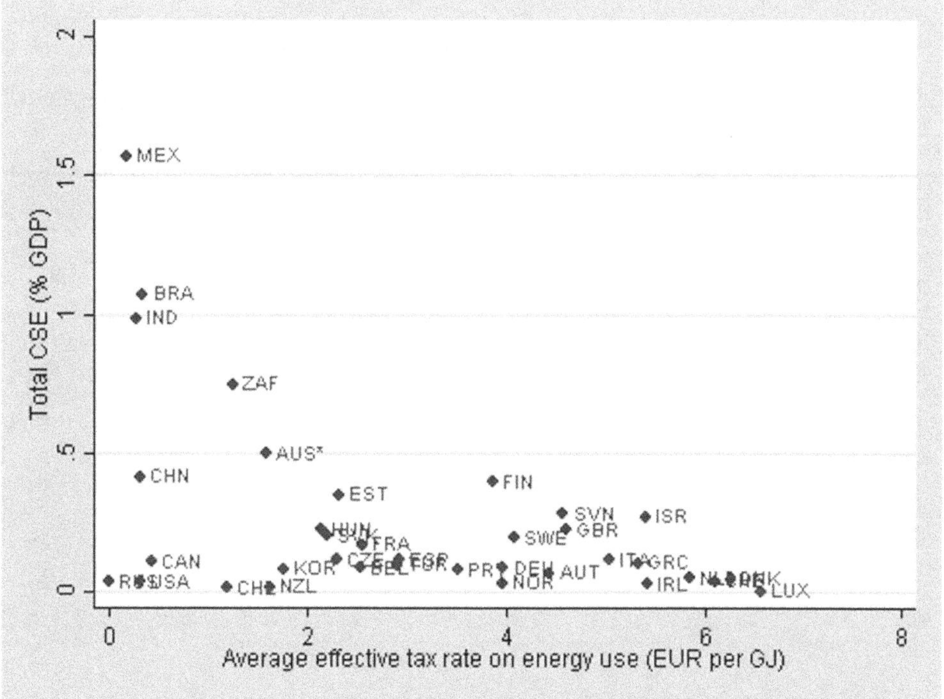

Notes: *The data for Australia include the country's large Fuel Tax Credits, which alone explain the relatively high ratio observed for that particular country. This measure serves to rebate some of the excise taxes that businesses pay on their purchases of fuel there. Data on average effective rates of tax on energy use come from OECD (2015b). Tax rates are as of 1 April 2012, except 1 July 2012 for Australia and Brazil and 4 April 2012 for South Africa. For that reason, the rates for Australia include the carbon tax that was subsequently repealed effective 1 July 2014. Rates for Canada, India, and the United States include federal taxes only.

3.4. Conclusions and policy implications: Paving the way for reform

The overall impression conveyed by the data compiled for this 2015 edition of the OECD *Inventory* is one of progress. Compared with the previous edition released in January 2013 (OECD, 2013b), for which the data stopped in 2011, total support for fossil fuels in OECD countries clearly exhibits a downward trend. With the new addition of estimates for a selection of partner economies (Brazil, China, India, Indonesia, the Russian Federation, and South Africa), this 2015 edition makes it possible to observe that a notable decline in support has also been underway in these countries since 2012. Underlying this decrease in support are two intertwined phenomena: the recent decline in international oil prices (Figure 3.5), an exogenous factor, and the actual reform efforts of several governments. This chapter has highlighted many such efforts, including the recent steps taken by Mexico, Indonesia, and India, three countries that have drastically reduced their support for the consumption of petroleum fuels.

Although progress is notable, the Inventory shows that there remains plenty of room for reform. The context is also not one for complacency. Global GHG emissions are still largely above the levels required for limiting average temperature increases. Recovery from the Great Recession of 2008-09 remains slow and difficult by historical standards. Fiscal positions continue posing a challenge to policy makers in many countries as they struggle to identify opportunities for cutting spending and raising more revenues, and this without adding to alarmingly high levels of unemployment. In this context, the reform of measures supporting fossil fuels appears more relevant than ever. Care should nevertheless be taken to ensure that reforms do not add to the plight of the poorest. Reforming support for fossil fuels will thus often form part of a broader strategy mobilising different parts of the government, including social assistance where necessary.

Figure 3.5. The evolution of international crude-oil prices, 2008-15

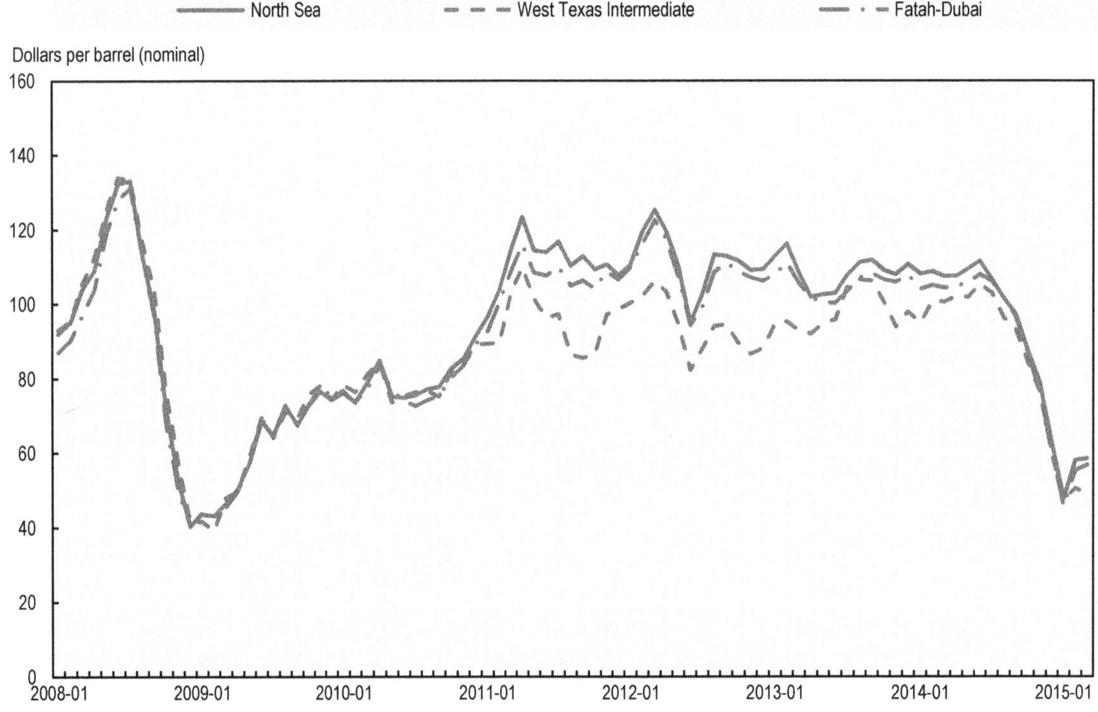

Source: IEA (2015b), *IEA Energy Prices and Taxes Statistics* (database).
DOI: http://dx.doi.org/10.1787/ene-pric-data-en.

In today's context, some countries may view support for the production of fossil fuels as a relatively easy way to increase future revenues (through higher royalties, resource taxes or severance taxes) and employment. It is indeed common for countries that are relatively well endowed with natural resources to fine-tune their tax system and adjust government take so as to improve the economics of particular projects and encourage more extraction of fossil fuels than would otherwise be the case. In normal times, this could be regarded as conventional practice, or at least acceptable practice, if only considerations of resource rent and energy security were involved. The times are, however, not normal, and efforts to curb GHG emissions worldwide remain insufficient to date. This therefore raises the question of the appropriateness of certain policies seeking to encourage the extraction of fossil fuels. Most policy discussions have so far centred on the consumption of fossil fuels but the time is likely ripe for starting a discussion on the production side too. It is particularly so as the low prices for hydrocarbons and coal that have prevailed in the first half of 2015 have strongly curtailed the revenues of extractive industries worldwide, which accentuates the pressures on governments to support fossil-fuel producers.

More generally, support measures were historically introduced for various reasons, each policy having its own *raison d'être*. Some were introduced to explicitly encourage the production or use of fossil fuels. Others were adopted with a very different purpose in mind. Either way, governments should periodically reassess those measures against their initial objectives and in light of today's changing economic and environmental landscape. Other better-targeted policy instruments likely exist and would offer suitable alternatives for meeting the stated policy objective(s). This is, for example, the case where measures seek to support the incomes of households by means of lower fuel taxes or direct energy subsidies. Given the objective of helping households, policies that directly support low incomes (e.g. redistribution through the normal income tax system or means-tested assistance) and those that improve the energy efficiency of buildings and appliances would likely do a better job than measures encouraging the consumption of energy.

Notes

1. Henceforth "China".
2. Figure A.1 in the Annex shows the composition of support by fuel and by indicator for each country.
3. See Chapter 2 for an explanation of the concept of formal or statutory incidence.
4. These policies include the Strategic Petroleum Reserve (1975), the Low-Income Home Energy Assistance Program (1981), the Alternative Fuels Production Credit (1986), the Expensing of Exploration and Development Costs (1986), and the Exception from Passive Loss Limitation (1986).
5. Data on the revenues countries derive from environmentally related taxes — which include taxes related to the use of energy, motor-vehicle taxes, and other environmental fees and levies (e.g. on waste and water use) — are regularly collected by the OECD and made available through the Organisation's Database of instruments used for environmental policy (www2.oecd.org/ecoinst/queries/).
6. Belgium's tax expenditure in relation to gasoil used in the residential sector, which uses as a benchmark the country's relatively high tax rate for diesel fuel used on roads, is similarly removed to improve comparability.
7. Those rates are the ones calculated for the companion publication *Taxing Energy Use 2015: OECD and Selected Partner Economies* (OECD, 2015b). See Box 2.3 in Chapter 2 for more details.

References

ADB (2011), *Assessment and Implications of Rationalizing and Phasing Out Fossil-Fuel Subsidies*, Technical Assistance Report, Project No. 45077-001, Asian Development Bank, July.

Albrizio, S., E. Botta, T. Koźluk and V. Zipperer (2014), "Do Environmental Policies Matter for Productivity Growth?: Insights from New Cross-Country Measures of Environmental Policies", *OECD Economics Department Working Papers*, No. 1176, OECD Publishing, Paris, DOI: http://dx.doi.org/10.1787/5jxrjncjrcxp-en

Ansar, A., B. Caldecott and J. Tilbury (2013), *Stranded assets and the fossil fuel divestment campaign: what does divestment mean for the valuation of fossil fuel assets?*, Stranded Assets Programme, 8 October, Smith School of Enterprise and the Environment, University of Oxford.

APEC (2009), *A New Growth Paradigm For a Connected Asia-Pacific in the 21st Century*, Statement by APEC Leaders, 14-15 November, Singapore.

Arlinghaus, J. (2015), "Impacts of Carbon Prices on Indicators of Competitiveness: A Review of Empirical Findings", *OECD Environment Working Papers*, No. 87, OECD Publishing, Paris. DOI: http://dx.doi.org/10.1787/5js37p21grzq-en

Boadway, Robin and Neil Bruce (1984), "A general proposition on the design of a neutral business tax", *Journal of Public Economics*, Vol. 24, No. 2, pp. 231-239.

Breisinger, C., W. Engelke and O. Ecker (2011), *Petroleum Subsidies in Yemen: Leveraging Reform for Development*, International Food Policy Research Institute, Development Strategy and Governance Division, IFPRI Discussion Paper No. 01071, Washington, D.C.

Bundesministerium der Finanzen (2013), *24. Subventionsbericht: Bericht der Bundesregierung über die Entwicklung der Finanzhilfen des Bundes und der Steuervergünstigungen für die Jahre 2011 bis 2014* (24th Subsidy Report: Report of the Federal Government on the Development of Federal Financial Assistance and Tax Expenditures for the Years 2011 to 2014), Federal Government of Germany, August, Berlin.

Burniaux, Jean-Marc, Jean Chateau and Jehan Sauvage (2011), "The Trade Effects of Phasing Out Fossil-Fuel Consumption Subsidies", *OECD Trade and Environment Working Papers*, No. 2011/05, OECD Publishing, Paris, DOI: http://dx.doi.org/10.1787/5kg6lql8wk7b-en

Butt, N., H.L. Beyer, J.R. Bennett, D. Biggs, R. Maggini, M. Mills, A.R. Renwick, L.M. Seabrook and H.P. Possingham (2013), "Biodiversity Risks from Fossil Fuel Extraction", *Science*, Vol. 342, 25 October, pp. 425-426.

Chateau, J., R. Dellink and E. Lanzi (2014), "An Overview of the OECD ENV-Linkages Model: Version 3", *OECD Environment Working Papers*, No. 65, OECD Publishing, Paris. DOI: http://dx.doi.org/10.1787/5jz2qck2b2vd-en

Durand-Lasserve, O., L. Campagnolo, J. Chateau and R. Dellink (2015), "Modelling of distributional impacts of energy subsidy reforms: an illustration with Indonesia", *OECD Environment Working Papers*, No. 86, OECD Publishing, Paris, DOI: http://dx.doi.org/10.1787/5js4k0scrqq5-en

CRS (2012), *Tax Expenditures: Compendium of Background Material on Individual Provisions*, Report prepared for the use of the Committee on the Budget by the Congressional Research Service, US Congress, December, Washington, D.C.

Dutch Safety Board (2015), *Earthquake risks in Groningen: An investigation into the role of the safety of citizens during the decision-making process on gas extraction (1959-2014)*, Summary, February, The Hague.

EIA (2011), *Direct Federal Financial Interventions and Subsidies in Energy in Fiscal Year 2010*, U.S. Energy Information Administration, U.S. Department of Energy, U.S. Government, July, Washington D.C.

Ellis, J. (2010), *The Effects of Fossil-Fuel Subsidy Reform: A review of modelling and empirical studies*, The Global Subsidies Initiative of the International Institute for Sustainable Development, Geneva, March.

European Commission (2014a), *European Resource Efficiency Platform (EREP): Manifesto & Policy Recommendations*, Directorate-General for the Environment, Brussels, March.

European Commission (2014b), *Enhancing comparability of data on estimated budgetary support and tax expenditures for fossil fuels*, Directorate-General for the Environment, Brussels, August.

Flues, F. and B. Johannes Lutz (2015), "Competitiveness Impacts of the German Electricity Tax", *OECD Environment Working Papers*, No. 88, OECD Publishing, Paris.
DOI: http://dx.doi.org/10.1787/5js0752mkzmv-en.

GAO (2005), *Government Performance and Accountability: Tax Expenditures Represent a Substantial Federal Commitment and Need to Be Re-examined*, GAO-05-690, US Government Accountability Office, September, Washington D.C.

Gauci, V., N.B. Dise, G.H. and M.E. Jenkins (2008), “Suppression of rice methane emission by sulfate deposition in simulated acid rain”, *Journal of Geophysical Research*, Vol. 113, No. G3.

G-20 (2009), *Leaders' Statement*, The Pittsburgh Summit, 24-25 September.

Harding, M. (2014), “The Diesel Differential: Differences in the Tax Treatment of Gasoline and Diesel for Road Use”, *OECD Taxation Working Papers*, No. 21, OECD Publishing, Paris,
DOI: http://dx.doi.org/10.1787/5jz14cd7hk6b-en

IADB (2013), *Documento de Cooperación Técnica: Subsidios en el sector energía y políticas de mitigación al cambio climático en América Latina y el Caribe* (Technical Co-operation Document: Subsidies to the energy sector and climate-change mitigation policies in Latin America and the Caribbean), RG-T2119, Inter-American Development Bank, August.

IEA (2015a), *Energy Technology Perspectives 2015*, International Energy Agency, OECD Publishing, Paris,
DOI: http://dx.doi.org/10.1787/energy_tech-2015-en.

IEA (2015b), *IEA Energy Prices and Taxes Statistics* (database).
DOI: http://dx.doi.org/10.1787/ene-pric-data-en.

IEA (2014a), *World Energy Outlook 2014*, International Energy Agency, OECD Publishing, Paris,
DOI: http://dx.doi.org/10.1787/weo-2014-en

IEA (2014b), *Energy Policies of IEA Countries: The United States 2014*, International Energy Agency, OECD Publishing, Paris,
DOI: http://dx.doi.org/10.1787/9789264211469-en

IEA (2013), *Technology Roadmap: Carbon Capture and Storage*, 2013 Edition, International Energy Agency, Paris.

IEA (1988), *Coal Prospects and Policies in IEA Countries: 1987 Review*, International Energy Agency, OECD Publishing, Paris.

IEA, OECD and Eurostat (2004), *Energy Statistics Manual*, OECD Publishing, Paris, DOI: http://dx.doi.org/10.1787/9789264033986-en

IMF (2013), *Energy subsidy reform: lessons and implications*, Edited by Benedict Clements, David Coady, Stefania Fabrizio, Sanjeev Gupta, Trevor Alleyne and Carlo Sdralevich, International Monetary Fund, Washington, D.C.

IMF (2012), *The* Delega Fiscale *and the Strategic Orientation of Tax Reform*, International Monetary Fund, Fiscal Affairs Department, IMF Country Report No. 12/280, October, Washington, D.C.

IPCC (2014), *Climate Change 2014: Synthesis Report*, Fifth Assessment Report of the Intergovernmental Panel on Climate Change, Geneva.

IRS (2014), *How to Depreciate Property*, Internal Revenue Service, Publication 946, 28 January, Department of the Treasury, US Government, Washington, D.C.

JCT (2014), *Estimates of Federal Tax Expenditures for Fiscal Years 2014-2018*, Report prepared for the House Committee on Ways and Means and the Senate Committee on Finance by the staff of the Joint Committee on Taxation, JCX-97-14, US Congress, 5 August, Washington, D.C.

IVM (2013), *Budgetary support and tax expenditures for fossil fuels: An inventory for six non-OECD EU countries*, IVM Institute for Environmental Studies, Report commissioned by the Directorate-General for the Environment of the European Commission, 15 January, Amsterdam.

Kojima, M. (2013), *Petroleum product pricing and complementary policies: Experience of 65 developing countries since 2009*, The World Bank, Sustainable Energy Department, Oil, Gas, and Mining Unit, Policy Research Working Paper No. 6396, April, Washington, D.C.

Kojima, M. and D. Koplow (2015), *Fossil Fuel Subsidies: Approaches and Valuation*, The World Bank, Energy and Extractives Global Practice Group, Policy Research Working Paper No. 7220, March, Washington, D.C.

Koplow, D. (2009), *Measuring Energy Subsidies Using the Price-Gap Approach: What does it leave out?*, Trade, Investment and Climate Change Series, August, International Institute for Sustainable Development, Winnipeg.

Koźluk, T. and V. Zipperer (2013), "Environmental Policies and Productivity Growth: A Critical Review of Empirical Findings", *OECD Economics Department Working Papers*, No. 1096, OECD Publishing, Paris, DOI: http://dx.doi.org/10.1787/5k3w725lhgf6-en

Lucas, D. (2015, forthcoming), *Inventory of Support for Fossil Fuels — Estimating the Support Component of Loan Guarantees and Preferential Loans*, OECD, Paris.

Lunden, L.P. and D. Fjaertoft (2014), *Government Support to Upstream Oil & Gas in Russia: How Subsidies Influence the Yamal LNG and Prirazlomnoe Projects*, The Global Subsidies Initiative of the International Institute for Sustainable Development, WWF Russia, July, Geneva-Oslo-Moscow.

Ministry of Finance of India (2015), *Receipt Budget 2015-2016*, Union Budget 2015-16, Central Government of India.

Ministry of Finance of the People's Republic of China (n.d.), *2013年全国财政决算* (2013 National Financial Accounts), http://yss.mof.gov.cn/2013qgczjs/, accessed 10 April 2015.

National Research Council (2013), *Effects of U.S. Tax Policy on Greenhouse Gas Emissions*, National Research Council, The National Academies Press, Washington, D.C.

OECD (2015a), *Aligning Policies for a Low-carbon Economy*, OECD Publishing, Paris. DOI: http://dx.doi.org/10.1787/9789264233294-en

OECD (2015b), *Taxing Energy Use 2015: OECD and Selected Partner Economies*, OECD Publishing, Paris. DOI: http://dx.doi.org/10.1787/9789264232334-en

OECD (2015c, forthcoming), *Climate Change Mitigation: Policies and Progress*, OECD Publishing, Paris.

OECD (2014a), *The Cost of Air Pollution: Health Impacts of Road Transport*, OECD Publishing, Paris, DOI: http://dx.doi.org/10.1787/9789264210448-en

OECD (2014b), *Agricultural Policy Monitoring and Evaluation 2014: OECD Countries*, OECD Publishing, Paris, DOI: http://dx.doi.org/10.1787/agr_pol-2014-en

OECD (2013a), *Effective Carbon Prices*, OECD Publishing, Paris, DOI: http://dx.doi.org/10.1787/9789264196964-en

OECD (2013b), *Inventory of Estimated Budgetary Support and Tax Expenditures for Fossil Fuels 2013*, OECD Publishing, Paris, DOI: http://dx.doi.org/10.1787/9789264187610-en

OECD (2012a), *OECD Economic Surveys: Indonesia 2012*, OECD Publishing, Paris, DOI: http://dx.doi.org/10.1787/eco_surveys-idn-2012-en

OECD (2012b), *Inventory of Estimated Budgetary Support and Tax Expenditures for Fossil Fuels*, OECD Publishing, Paris, DOI: http://dx.doi.org/10.1787/9789264128736-en

OECD (2010a), *Tax Expenditures in OECD Countries*, OECD Publishing, Paris. DOI: http://dx.doi.org/10.1787/9789264076907-en

OECD (2010b), *OECD's Producer Support Estimate and Related Indicators of Agricultural Support: Concepts, Calculations, Interpretation and Use (The PSE Manual)*, Trade and Agriculture Directorate, September, Paris, PSE Manual - OECD

OECD (2009), *Declaration on Green Growth*, Adopted at the Meeting of the Council at Ministerial Level on 25 June 2009, [C/MIN(2009)5/ADD1/FINAL].

OECD (2003), *Environmentally Harmful Subsidies: Policy Issues and Challenges*, OECD Publishing, Paris, DOI: http://dx.doi.org/10.1787/9789264104495-en

OECD (2001), *OECD Environmental Strategy for the First Decade of the 21st Century*, Adopted by OECD Environment Ministers, 16 May, Paris.

OMB (2015), *Cuts, Consolidations, and Savings*, Budget of the United States Government, Fiscal Year 2016, Office of Management and Budget, US Government, Washington, D.C.

Parry, I., D. Heine, E. Lis and S. Li (2014), *Getting Energy Price Right: From Principle to Practice*, International Monetary Fund, Washington, D.C.

Sambijantoro, S. (2015), "Jokowi's budget aims for economic liftoff", *The Jakarta Post*, 13 February, Jakarta.

Sauvage, J. (2014), "The Stringency of Environmental Regulations and Trade in Environmental Goods", *OECD Trade and Environment Working Papers*, No. 2014/03, OECD Publishing, Paris, DOI: http://dx.doi.org/10.1787/5jxrjn7xsnmq-en

Singapore Government (2015), *Customs Act (Chapter 70)*, Singapore Statutes Online, http://statutes.agc.gov.sg/aol/search/display/view.w3p;page=0;query=DocId%3A%22f3e1a9ec-037e-450c-a9a5-343bf5eab23e%22%20Status%3Ainforce%20Depth%3A0;rec=0, accessed 2 May 2015.

United Nations (2014), *Open Working Group proposal for Sustainable Development Goals*, New York.

Zero CO_2 database (n.d.), www.zeroco2.no/projects, accessed 11 May 2015.

Annex A

Table A.1. The IEA's classification of fossil fuels

Broad category	IEA Short name	IEA full name
Solid fuels	ANTCOAL	Anthracite
	BITCOAL	Other bituminous coal
	BKB	BKB
	BROWN	Brown coal (if no detail)
	COALTAR	Coal tar
	COKCOAL	Coking coal
	GASCOKE	Gas coke
	HARDCOAL	Hard coal (if no detail)
	LIGNITE	Lignite
	OILSHALE	Oil shale and oil sands
	OVENCOKE	Coke oven coke
	PATFUEL	Patent fuel
	PEAT	Peat
	SUBCOAL	Sub-bituminous coal
Liquid fuels and associated products	ADDITIVE	Additives and blending components
	AVGAS	Aviation gasoline
	BITUMEN	Bitumen
	CRNGFEED	Crude, NGL, or feedstocks (if no detail)
	CRUDEOIL	Crude oil
	ETHANE	Ethane
	JETGAS	Gasoline type jet fuel
	LPG	Liquefied petroleum gases (LPG)
	LUBRIC	Lubricants
	NAPHTHA	Naphtha
	NGL	Natural gas liquids
	NONBIODIES	Gasoil or diesel oil, excl. biofuels
	NONBIOGASO	Motor gasoline excl. biofuels
	NONBIOJETK	Kerosene type jet fuel excl. biofuels
	NONCRUDE	Other hydrocarbons
	ONONSPEC	Other oil products
	OTHKERO	Other kerosene
	PARWAX	Paraffin waxes
	PETCOKE	Petroleum coke
	REFFEEDS	Refinery feedstocks
	RESFUEL	Fuel oil
	WHITESP	White spirit & SBP
Gaseous fuels	BLFURGS	Blast furnace gas
	COKEOVGS	Coke oven gas
	GASWKSGS	Gas works gas
	NATGAS	Natural gas
	REFINGAS	Refinery gas

Source: Adapted from the IEA, wds.iea.org/WDS/tableviewer/document.aspx?FileId=1496.

Table A.2. MFN tariffs applied by OECD countries and partner economies on their imports of hydrocarbon fuels (as of January 2015)

Country	Crude oil and liquid petroleum products							Gaseous hydrocarbons		
	Crude oil	Motor gasoline	Aviation spirit	Kerosene	Jet fuel, kerosene-based	Diesel	Heavy fuel oil	LNG	LPG	Gaseous natural gas
HS code	2709	2710.11 ex	2710.11 ex	2710.19 ex	2710.19 ex	2710.19 ex	2710.19 ex	2711.11	2711.12	2711.21
Australia[1]	0%	0%	0%	0%	0%	0%	0%	0%	0%	0%
Brazil	0%	0%	0%	0%	0%	0%	0%	0%	0%	0%
Canada	0%	0%	0%	0%	0%	0%	0%	0%	0-12.5%	0%
Chile	6%	6%	6%	6%	6%	6%	6%	6%	6%	6%
China, People's Republic of	0%	0-9%	0-9%	0-9%	0%	0-6%	1%	0%	1-11%	0-6%
Iceland	0%	0%	0%	0%	0%	0%	0%	0%	0%	0%
India	0%	2.5-5%	0%	5%	5%	5%	5%	5%	2.5-5%	5%
Indonesia	0%	0%	0%	0%	0%	0%	0%	5%	5%	5%
Israel[2]	0%	0%	0%	0%	0%	0%	0%	0%	0%	0%
European Union	0%	4.7%	4.7%	4.7%	4.7%	0-3.5%	3.5%	0%	0-8%	0%
Japan	0%	JPY 934/kL	JPY 934/kL	0-3%	JPY 346/kL	JPY 750/kL	JPY 0-459/kL	0%	0%	4.1%
Korea	3%	5%	5%	3-5%	3-5%	5%	3-5%	2%	2%	2%
Mexico	0%	0%	0%	0%	0%	0%	0%	0%	0%	0%
New Zealand	0%	0%	0%	0-5%	0-5%	0%	0%	0%	NZD 0.104/L	NZD 3.17/GJ
Norway	0%	0%	0%	0%	0%	0%	0%	0%	0%	0%
The Russian Federation	0-5%	0-5%	0%	5%	5%	5%	5%	5%	5%	5%
South Africa	0%	ZAR 0-0.091/L	0%	ZAR 0.183/L	0%	ZAR 0.183/L	0%	0%	0%	0-15%
Switzerland	0%	0%	0%	0%	0%	0%	0%	0%	0%	0%
Turkey	0%	4.7%	4.7%	4.7%	4.7%	0-3.5%	3.5%	0%	0-8%	0%
United States	USD 0.0525-0.21/bbl	USD 0.525-1.05 / bbl	USD 0.0525-1.05/bbl	USD 0.105-0.21/bbl	USD 0.525-1.05/ bbl	USD 0.105-0.525/bbl	USD 0.0525-0.21/bbl	0%	0%	0%

1. Australia applies excise duties at the point of import, and lists these duties in its tariff schedule. Since these (AUD 0.392 per litre for motor gasoline, kerosene, diesel and heavy fuel oil; AUD 0.03556 per litre for aviation spirit and jet fuel; AUD 0.128 per litre for LPG; and AUD 0.268 per kg for LNG and gaseous natural gas) are the same as the normal excise duty applied to domestically produced fuels, the tariffs here are listed as zero.

2. The statistical data for Israel are supplied by and under the responsibility of the relevant Israeli authorities. The use of such data by the OECD is without prejudice to the status of the Golan Heights, East Jerusalem and Israeli settlements in the West Bank under the terms of international law.

Source: European Union: Business Link (www.businesslink.gov.uk/bdotg/action/tariff); all other countries: European Commission, Market Access Database (http://madb.europa.eu/madb/indexPubli.htm).

Table A.3. MFN tariffs applied by OECD countries and partner economies on their imports of solid fossil fuels (as of January 2015)

Country	Hard coal				Lignite		Peat	Coke and semi-coke
	Anthracite	Bituminous coal	Other	Briquettes of hard coal	Non-agglomerated	Agglomerated		or coal, lignite or peat
HS code:	2701.11	2701.12	2701.19	2701.20	2702.10	2702.20	2703	2704
Australia	0%	0%	0%	0%	0%	0%	0%	0%
Brazil	0%	0%	0%	0%	0%	0%	0%	0%
Canada	0%	0%	0%	0%	0%	0%	6.5%	0%
Chile	6%	6%	6%	6%	6%	6%	6%	6%
China	0%	0%	0%	0%	3%	3%	5%	5%
Iceland	0%	0%	0%	0%	0%	0%	0%	0%
India	5%	5%	5%	5%	5%	5%	5%	5%
Indonesia	5%	5%	5%	5%	5%	5%	5%	5%
Israel[1]	0%	0%	0%	0%	0%	0%	6%	0%
European Union	0%	0%	0%	0%	0%	0%	0%	0%
Japan	0%	0%	0%	3.9%	0%	0%	0%	3.2%
Korea	0%	0%	0%	1%	1%	1%	1%	3%
Mexico	0%	0%	0%	0%	0%	0%	0%	0%
New Zealand	0%	0%	0%	0%	0%	0%	0%	0%
Norway	0%	0%	0%	0%	0%	0%	0%	0%
Russia	5%	0-5%	5%	5%	5%	5%	5%	0%
South Africa	0%	0%	0%	0%	0%	0%	0%	0%
Switzerland	CHF 0.80/ tonne	CHF 0.80/ tonne	CHF 0.80/ tonne	CHF 0.80/ tonne	CHF 0.80/ tonne	CHF 0.80/ tonne	CHF 0.80/ tonne	CHF 0.80/ tonne
Turkey	0%	0%	0%	0%	0%	0%	0%	0%
United States	0%	0%	0%	0%	0%	0%	0%	0%

1. The statistical data for Israel are supplied by and under the responsibility of the relevant Israeli authorities. The use of such data by the OECD is without prejudice to the status of the Golan Heights, East Jerusalem and Israeli settlements in the West Bank under the terms of international law.

Sources: European Union: Business Link (*www.businesslink.gov.uk/bdotg/action/tariff*); all other countries: European Commission, Market Access Database (*madb.europa.eu/mkaccdb2/indexPubli.htm*). The identification of support measures was conducted mainly through searches of official government documents and web sites. In a few cases, unpublished data were requested from, and furnished by, OECD governments.

Table A.4. Matrix of support measures with examples

Transfer Mechanism (how a transfer is created)	Statutory or Formal Incidence (to whom and what a transfer is first given)								
	Production							Direct consumption	
	Output returns	Enterprise Income	Cost of intermediate inputs	Costs of production factors					
				Labour	Land and natural resources	Capital	Knowledge	Unit cost of consumption	Household or enterprise income
Direct transfer of funds	Output bounty or deficiency payment	Operating grant	Input-price subsidy	Wage subsidy	Capital grant linked to acquisition of land	Capital grant linked to capital	Government R&D	Unit subsidy	Government-subsidized life-line electricity rate
Tax revenue foregone	Production tax credit	Reduced rate of income tax	Reduction in excise tax on input	Reduction in social charges (payroll taxes)	Property-tax reduction or exemption	Investment tax credit	Tax credit for private R&D	VAT or excise-tax concession on fuel	Tax deduction related to energy purchases that exceed given share of income
Other government revenue foregone			Under-pricing of a government good or service		Under-pricing of access to government land or natural resources; Reduction in resource royalty or extraction tax		Government transfer of intellectual property right	Under-pricing of access to a natural resource harvested by final consumer	
Transfer of risk to government	Government buffer stock	Third-party liability limit for producers	Provision of security (e.g., military protection of supply lines)	Assumption of occupational health and accident liabilities	Credit guarantee linked to acquisition of land	Credit guarantee linked to capital		Price-triggered subsidy	Means-tested cold-weather grant
Induced transfers	Import tariff or export subsidy	Monopoly concession	Monopsony concession; export restriction	Wage control	Land-use control	Credit control (sector-specific)	Deviations from standard IPR rules	Regulated price; cross subsidy	Mandated life-line electricity rate

Figure A.1. Composition of total support by fuel (left) and indicator (right)

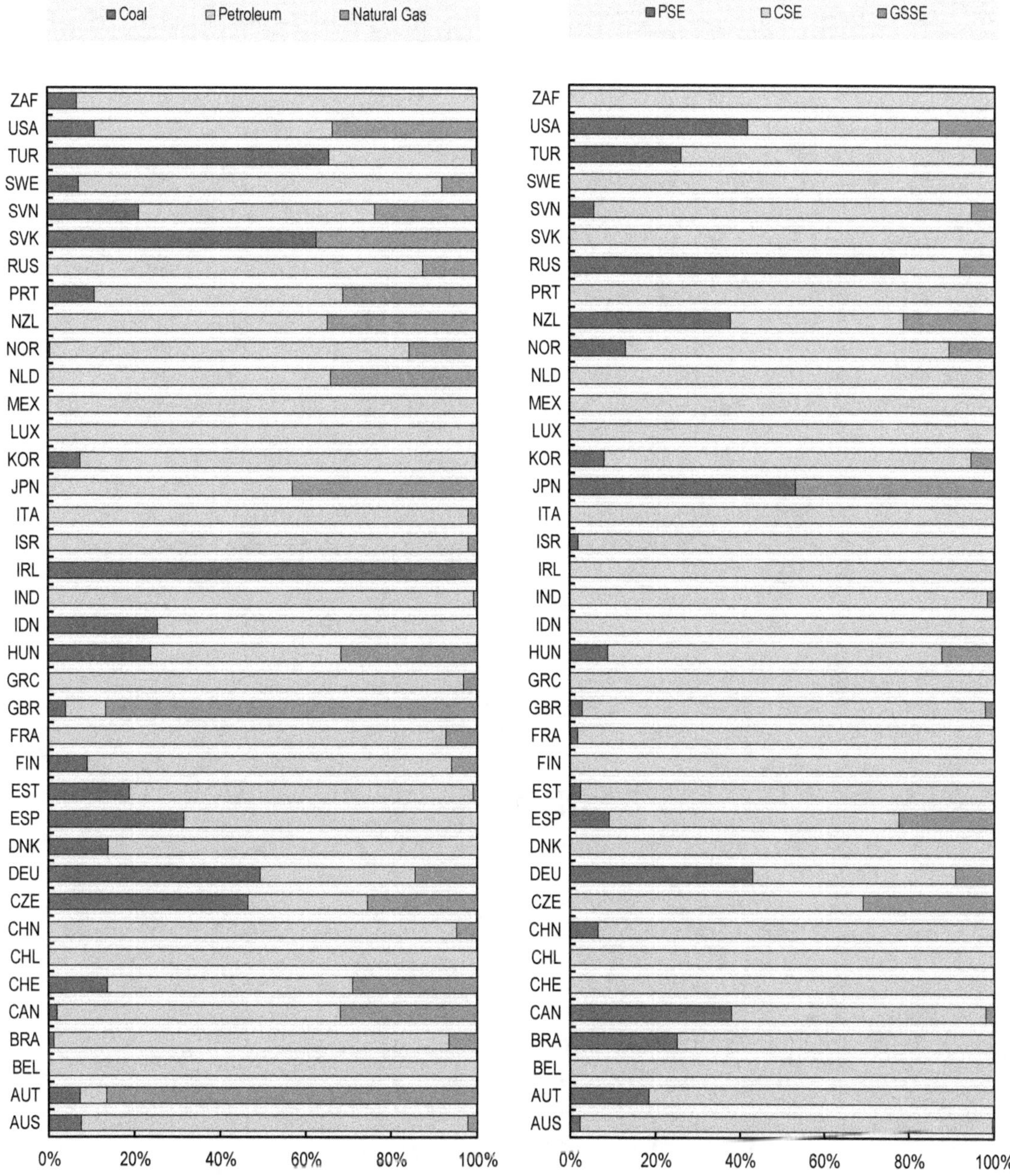

ORGANISATION FOR ECONOMIC CO-OPERATION AND DEVELOPMENT

The OECD is a unique forum where governments work together to address the economic, social and environmental challenges of globalisation. The OECD is also at the forefront of efforts to understand and to help governments respond to new developments and concerns, such as corporate governance, the information economy and the challenges of an ageing population. The Organisation provides a setting where governments can compare policy experiences, seek answers to common problems, identify good practice and work to co-ordinate domestic and international policies.

The OECD member countries are: Australia, Austria, Belgium, Canada, Chile, the Czech Republic, Denmark, Estonia, Finland, France, Germany, Greece, Hungary, Iceland, Ireland, Israel, Italy, Japan, Korea, Luxembourg, Mexico, the Netherlands, New Zealand, Norway, Poland, Portugal, the Slovak Republic, Slovenia, Spain, Sweden, Switzerland, Turkey, the United Kingdom and the United States. The European Union takes part in the work of the OECD.

OECD Publishing disseminates widely the results of the Organisation's statistics gathering and research on economic, social and environmental issues, as well as the conventions, guidelines and standards agreed by its members.

OECD PUBLISHING, 2, rue André-Pascal, 75775 PARIS CEDEX 16
(22 2015 01 1 P) ISBN 978-92-64-23960-9 – 2015-02

www.ingramcontent.com/pod-product-compliance
Lightning Source LLC
LaVergne TN
LVHW081423110826
845149LV00010B/1843

* 9 7 8 9 2 6 4 2 3 9 6 0 9 *